VASCULAR PLANTS OF THE
South Sound Prairies

FIRST EDITION

EDITED BY FREDERICA BOWCUTT
AND SARAH HAMMAN

The Evergreen State College Press
Olympia, Washington

Profits from the sale of this book will be donated to support
prairie restoration and the Evergreen Herbarium.

Copyright © 2016 by The Evergreen State College Press
All rights reserved.

Printed and bound in the United States of America
Design: Jackie Argueta and Natalie Hammerquist
First Edition

The Evergreen State College Press
www.evergreen.edu/press/

Library of Congress Cataloging-in-Publication Data
Vascular Plants of the South Sound Prairies/ Frederica Bowcutt and Sarah Hamman, editors. - 1st ed.
pages cm.
Includes bibliographical references and index.
ISBN 978-0-9974158-0-3
1. Prairies. 2. Prairie - Floristics. 3. Prairie - Ecology. 4. Prairie - Restoration Ecology. 5. Ethnobotany. 6. Pacific Northwest - Cultural Landscapes. 7. Pacific Coast (U.S.) - Cultural Landscapes. 8. Human Ecology - Washington

Front cover photo of showy fleabane (*Erigeron speciosus*) taken by Rod Gilbert at his home prairie garden. Front cover photo of common camas (*Camassia quamash*) and back cover photo of deltoid balsamroot (*Balsamorhiza deltoidea*) taken at Glacial Heritage Preserve by Frederica Bowcutt.

Contributors

INTRODUCTION	Frederica Bowcutt
STUDY AREA DESCRIPTION	Peter Dunwiddie, Ed Alverson, Amanda Stanley, Rod Gilbert, Scott Pearson, Dave Hays, Joe Arnett, Eric Delvin, Dan Grosboll, Caroline Marschner
CLIMATE	Lisa Hintz
GEOLOGY	Pat Pringle
VEGETATION	Joe Bettis
SENSITIVE SPECIES	Lisa Hintz
RESTORATION ECOLOGY	Sarah Hamman
HISTORY OF WHITE SETTLEMENT	Lisa Hintz
HOLISTIC RESTORATION	Rose Edwards
PLANT DESCRIPTIONS	Kathleen McSorley, Morgan Cheyney, Saff Killingsworth, Meg Krug and others
ILLUSTRATORS	Joe Bettis, Yianna Bekris, Frederica Bowcutt, Emily Driskill, Natalie Hammerquist, Lisa Hintz, Lily Hynson, Krista Koller, Meg Krug, Jordan Marlor, Callie Martin, Irene Matsuoka, Kathleen McSorley, Megan Porter, Stella Rose and Brita Zeiler
SPECIES LIST OF VOUCHER SPECIMENS (APPENDIX A)	Stella Rose, Emily Driskill, Spencer Rogers and Mike Hicks

Contents

Acknowledgements	vii
Introduction	1
Study Area Description	5
Climate	7
Geology	13
Vegetation	21
Sensitive Species	27
Restoration Ecology	37
History of White Settlement	43
Holistic Restoration	57
Plant Illustrations and Descriptions	59
List of Illustrated Plants	60
Appendix A: List of Prairie Voucher Specimens in The Evergreen State College Herbarium	130
Appendix B: Illustrators' Contributions	136
Glossary	139
Literature Cited	146
Index	160

Acknowledgements

This floristic study is built on specimens housed at the herbarium of The Evergreen State College in Olympia, Washington. In spring 2003, undergraduates Anna Constance, Vinson Doyle, Cody Hinchcliff and Rebecca Sheedy produced a vouchered floristic study of Glacial Heritage County Preserve. These students were in a year-long program I co-taught with artist Lucia Harrison called Picturing Plants. Evergreen Natural History Museum Fellow Heron Brae assisted Masters of Environmental Studies (MES) students with collecting voucher herbarium specimens in my spring 2009 Floristic Research Methods course. Students involved in this effort included Autumn Pickett, Natalie Pyrooz, Tracey Scalici, Taj Schade, Megan Treasure and Jill Zarzeczny (formerly Jill Politsch). Thanks to Angel Przybylowicz (formerly Angel Lombardi) for her donation of specimens from Fort Lewis. Also I am grateful to my late colleague and emeritus faculty member, Al Wiedemann, for starting the herbarium at Evergreen.

Contributors to the text include Joe Bettis, Morgan Cheyney, Peter Dunwiddie, Rose Edwards, Sarah Hamman, Lisa Hintz, Meg Krug, Kathleen McSorley, Pat Pringle and others. Stella Rose, Emily Driskill, Mike Hicks and Spencer Rogers compiled the vouchered species list based on herbarium specimens housed at Evergreen's herbarium. Yianna Bekris, Nikolai Starzak and Sofia Vasconi compiled the glossary. Joe Arnett, David Giblin, Saff Killingsworth, Kathleen McSorley, Stella Rose and especially Lisa Hintz provided copy editing and proofing assistance. Sixteen illustrators contributed to this project thus far. Undergraduate illustrators include: Yianna Bekris, Joe Bettis, Emily Driskill, Natalie Hammerquist, Lisa Hintz, Lily Hynson, Krista Koller, Meg Krug, Jordan Marlor, Callie Martin, Irene Matsuoka, Kathleen McSorley, Megan Porter, Stella Rose and Brita Zeiler. Two of these students, Jordan Marlor and Stella Rose, were trained in botanical illustration by my colleague Ruth Hayes. The remaining illustrators learned their craft in one of my field plant taxonomy focused programs. Hannah Anderson and Adam Martin of the Center for Natural Lands Management created the study area and geology maps, respectively. Rod Gilbert donated the photograph of showy fleabane for the cover. Undergraduate students Natalie Hammerquist produced the original graphic layout and Jackie Argueta completed the final design.

Multiple organizations and government agencies have facilitated our work by giving us permission to access their land and collect specimens. These include Thurston County, U.S. Department of Defense's Joint Base Lewis-McChord, Wolf Haven International, Washington Department of Fish and Wildlife, and Washington Department of Natural Resources. The Washington Native Plant Society provided financial assistance with our printing costs. Gorham Printing of Centralia, Washington printed and bound the final product.

For their support of this effort, I thank David Giblin who is the Herbarium Collection Manager at the University of Washington, Ben Legler and Peter Dunwiddie also at the University of Washington, Joint Base Lewis-McChord botanist Rod Gilbert, Eric Delvin with The Nature Conservancy, and Pat Dunn and Sarah Hamman, both with the Center for Natural Lands Management. Peter Dunwiddie generously gave us permission to use a slightly revised version of a coauthored study area description previously published in an article on the floristics of the south Puget Sound prairies.

This work would not have been possible without the support of The Evergreen State College. The college funded a sabbatical for me to complete focused work on this project in spring of 2015. In their roles as editorial board members of The Evergreen State College Press, Steven Hendricks and Brian Walter helped with the final file preparation and proofing, respectively. The administration and the Facilities Department invested in a state-of-the-art herbarium with climate control which enables our ongoing floristic research using non-toxic curation methods. Information technology experts on campus, particularly Amy Greene and David Geeraerts, made our web presence possible and helped with ongoing efforts to make our herbarium searchable online.

For updates on this field guide and other floristic research at Evergreen's herbarium, visit: blogs.evergreen.edu/naturalhistory. This field guide can be purchased from the Greener Store at: tescbookstore.com. Please contact us if you would like to donate herbarium specimens for this continuing effort. Suggested changes for the second edition are also welcomed and can be sent to bowcuttf@evergreen.edu or the address below.

Frederica Bowcutt, Director
Evergreen Natural History Museum
The Evergreen State College
Olympia, Washington 98505

VASCULAR PLANTS OF THE SOUTH SOUND PRAIRIES

FIRST EDITION

Introduction

Frederica Bowcutt

This field guide for the vascular plants of the south Puget Sound prairie-oak ecosystems is a work in progress involving over forty students from The Evergreen State College, as well as scientists from Evergreen, Centralia College, and the Center for Natural Lands Management. Through a collaborative, community-based effort that began in 2003, we have produced an illustrated guide to nearly 150 vascular plants coupled with text on the natural and cultural history of the glacial outwash prairies and their associated oak woodlands from Tacoma to Rochester, Washington. It includes descriptions of climate, geology, vegetation, sensitive species, the history of white settlement, and restoration ecology, plus a list of the voucher specimens that are maintained at the Evergreen Herbarium. Some of the species from wetter prairies on richer sites and along the forest edge are also included, but this was not our focus. We will update the nomenclature and family treatments in the second edition after the University of Washington's herbarium publishes the updated regional flora, which is expected within a few years. In addition to increasing the number of illustrated taxa, the voucher list in Appendix A will be expanded significantly. We also aspire to add a new section on Indigenous traditional ecological knowledge including burning practices. And we intend to convert the in-text citations to endnotes.

Given the growing interest in south Puget Sound prairie-oak ecosystems, this field guide fills a need among scientists, policymakers, and the educated public. Much floristic research on the prairies has been conducted, including numerous checklists of vascular plants. However, much of this work has not been paired with voucher specimens deposited in a herbarium that are made available for collective study. Also, much of this information is unpublished and difficult to access. None of the lists are illustrated. In addition, the illustrated field guides for the area, such as Pojar and MacKinnon's *Plants of the Pacific Northwest Coast*, focus primarily on plants of the coniferous forests. To identify prairie plants currently, the technical dichotomous keys in Hitchcock and Cronquist's *Flora of the Pacific Northwest* are the most commonly used resource. This regional flora is not easily used by beginners; nor does it generally allow for quick identification even among those familiar with using it. In addition, because it was published in 1973, both the nomenclature and classification are outdated.

Our aspiration is to produce an easy-to-use field guide to these threatened ecosystems to assist with important restoration and conservation work. The guide will support numerous field crews who monitor ongoing prairie and oak woodland restoration work each year. It will also aid the work of volunteers who collect native plant seeds, remove invasive species, and organize Prairie Appreciation Day, among other tasks. Many of these generous souls operate without formal training in botany but are eager to learn. Volunteer organizations like the Friends of Puget Prairies (FOPP) team with non-profit organizations and various government agencies to foster local prairie and oak wellness. In addition to aiding restoration work, we hope the field guide will also invite residents of the south Salish Sea (aka Puget Sound) region to expand their knowledge of these ecologically fascinating and biodiverse ecosystems.

This field guide demonstrates the capacity of college students to make significant contributions to their community through citizen science. Floristics is a branch of plant taxonomy focused on documenting the flora of various regions in the world. In tropical areas where biodiversity is being lost at an alarming rate, biologists have trained members of local communities to be research assistants. Parataxonomy training has led to new professional opportunities for under-represented populations, including Indigenous peoples. This project builds on this model by involving students in documenting plant diversity. Both undergraduate and graduate students of mine have contributed hundreds of herbarium specimens for the project since 2003 as a part of various academic programs and independent study contracts at The Evergreen State College. The illustration work began in spring 2011 when the first draft of this book was compiled as a part of my Field Plant Taxonomy program. Students in interdisciplinary programs that incorporate both botany and scientific illustration generated the pen and ink drawings.

Through herbarium-based research, I teach students how to use information technology to increase access beyond the academy. Images of our prairie voucher specimens are available through a website maintained by the Consortium of Pacific Northwest Herbaria. Historically, herbaria and botanical gardens have aided imperialist and colonial agendas. Botanists have served as scouts for natural resources to exploit at the expense of Indigenous peoples. Thus information sharing is an important response to this critique. It facilitates collaboration with local communities including tribes, public institutions, and non-profit organizations. In addition to making our voucher specimens available online, the field guide is paired with a website created by undergraduate students at Evergreen. It features Wikipedia-style pages for over 100 vascular plants with photographs generously donated by Rod Gilbert, Ben Legler, and others. Lisa Hintz provided photographs taken with an automontage microscope of the magnified seeds of most of the plants featured. These photographs aid in the seed identification work required for propagating native species for restoration. To access this wiki go to: wikis.evergreen.edu/pugetprairieplants

The technical expertise of how to foster these special places in our conifer-dominated region has grown significantly over the last few decades. Numerous animals and plants that depend on these ecosystems are threatened by the radical reduction in acreage since white settlement began in the mid-1800s. As a result, the Endangered Species Act drives much of the current prairie and oak woodland restoration work in our region. Species conservation work is needed. However, so is greater historical understanding of these cultural landscapes, which have depended on human intervention for thousands of years to thrive. Culturally sensitive ecological restoration is needed that honors the roles of Native Americans in the past to foster prairie and oak wellness. From the standpoint of social justice and environmental sustainability, it is also important to support contemporary tribal efforts to revitalize Indigenous food systems while honoring the long history of Native people tending these cultural landscapes.

Study Area Description

Peter Dunwiddie, Ed Alverson, Amanda Stanley, Rod Gilbert, Scott Pearson, Dave Hays, Joe Arnett, Eric Delvin, Dan Grosboll and Caroline Marschner

The largest remnants of upland prairie in western Washington are found in the southern Puget Sound region, where conservation efforts have been underway for nearly 40 years. Most of these sites are now protected, and many are under active management to control invasive species and restore degraded areas. The study area for this field guide includes 16 prairie sites with reasonably high native vascular plant species diversity (Fig. 1). These include all of the major remaining prairies, as well as several smaller fragments that still retain many of their native species. All probably were regularly burned by Native Americans prior to the mid-1800s and likely received some level of livestock grazing in the 19th and 20th centuries, although these histories are difficult to reconstruct in any detail.

Half of the study sites, including Johnson, Marion, 13th Division, 91st Division, 7S, Upper Weir, Lower Weir, and South Weir, are prairies on the Joint Base Lewis-McChord (JBLM), established in 1917. Although neglected after World War I, JBLM has been an active installation administered by the U.S. Department of Defense since the late 1930s. Military training occurs on most of the prairies, with the exception of South Weir. Fires are frequently ignited as a result of training exercises, particularly in the 91st Division prairie. The remaining study sites occur on non-military lands. Rocky Prairie and Mima Mounds are Natural Area Preserves managed by the Washington Department of Natural Resources.

Rocky Prairie probably has been little disturbed in many decades; Mima Mounds was grazed by livestock until the 1960s, but has been largely undisturbed since then except for restoration activities to control invasive grasses, shrubs, and trees. Scatter Creek and the recently acquired West Rocky Prairie are managed by the Washington Department of Fish and Wildlife for a variety of resource interests; neither has been grazed or burned in at least several decades. Glacial Heritage (officially the Black River-Mima Mounds Glacial Heritage Preserve) is owned by Thurston County, with restoration actions cooperatively managed originally with The Nature Conservancy and more recently with the Center for Natural Lands Management.

Fire management has been returned to portions of Glacial Heritage during the last 12 years, but this site has received little other use in several decades. Tenalquot Prairie is immediately adjacent to South Weir on Fort Lewis, but is private land outside the military reservation. It was very lightly grazed until 2005, when it was acquired by The Nature Conservancy. Finally, Wolf Haven is a privately owned wildlife rehabilitation center that includes a small, lightly-used area of native prairie habitat.

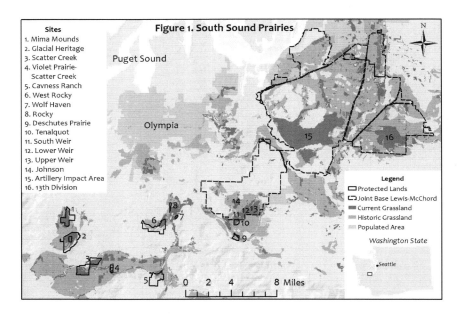

Figure 1. Map of the study area.

Climate

Lisa Hintz

Glaciation and Geology

The climatic event known as the Frasier Glaciation occurred from 25,000-10,000 years ago and most directly affected the region of the Puget Lowland (or Puget Sound basin) when it reached its maximum extent roughly 15,000-13,500 years ago (Barnosky 1985, 264; Whitlock 1992, 5). Although the glaciation of the Puget Lowland by the Puget Lobe of the Vashon ice sheet extended to just south of Olympia, Washington, the land beyond the glaciation was also affected. For more detail, see the section on geology.

Warming of the global climate contributed to the melting of the glacier, shaping the land beyond the limits of the ice by the sheer volume and force of meltwater drainage to the Pacific Ocean. The known effects in our region of these catastrophic events include: erosion and scouring of surface soils, degradation of soil fertility, deposition of gravel and silt, depressions with poor drainage (primarily around the southern edge of the glacier's final extent), and the carving of drainages into the landscape by meltwater (Franklin and Dyrness 1973, 17; Kruckeberg 1991, 288).

Climate and Floristics

Coarse details of the Earth's climate history are primarily reconstructed by examination of the layers of sediment found in long-existing lakebeds, bogs, and fens (Barnosky 1985, 263). Core samples of these sediment layers contain fossilized pollen and plant macro-fossils that indicate shifts in vegetation occurring over long time periods. These shifts include aspects of plant community composition such as diversity, abundance, and distribution (Barnosky 1985, 263; Whitlock 1992, 1). Changes in plant communities indicate transformations of local climate patterns. We can deduce what past climate conditions may have been like in areas where scientists have exhumed and analyzed core samples based on our current ecological knowledge of species-specific ranges and environmental limitations.

The Pollen Record

Much of what we can construct of climate trends in the Puget Trough of the past 20,000 years is based on sediment cores taken from areas that were under and/or adjacent to the Vashon ice sheet. Many important studies detail the fossil pollen layers that are held within these core samples (Hansen 1938 and 1947; Davis 1973; Barnosky 1981, 1984, and 1985; Tsukada, Sugita and

Hibbert 1981; Leopold, et al. 1982; Tsukada and Sugita 1982; Cwynar 1987; Dunwiddie 1987). With these studies, extensive vegetation and climate frameworks have been produced.

As organic and inorganic material moves throughout the environment, some inevitably lands on the surfaces of bodies of water, sinks to the bottom, and contributes to layers of sediment over time. Occasionally, material in these layers is preserved and fossilized. Paleoecologists remove core samples to be analyzed in the laboratory, which include fossilized pollen (often identifiable to the genus or species level), plant macro-fossils, volcanic ash, and charcoal fragments. Thus, sediment cores allow a chronological reading of some environmental components present throughout a time scale of thousands of years (Hansen 1947, 5). Timelines have been constructed using radio-carbon dating in conjunction with the positions of volcanic ash layers of known eruptions of Mount St. Helens, Glacier Peak, and Mount Mazama (the remnant of which is now known as Crater Lake) (Whitlock 1992, 4).

There are limitations to the data on the relative abundance and/or occurrence of species within the sediment core profile that must be taken into consideration. For example, some taxa generally produce less pollen than others; the shape and size of pollen varies extensively and this affects its degree of transport throughout the environment (e.g., taxa adapted to wind pollination are better represented in the fossil strata) (Hansen 1947, 6); and the potential of various pollen types to be preserved in sediment is variable. However, as postglacial vegetation ecologist Henry Hansen explained in his 1947 article "Postglacial Forest Succession, Climate, and Chronology in the Pacific Northwest" (7), "the degree of error is somewhat reduced by the magnitude of the time and space involved." By means of the science of paleoecology, we infer the bulk of our understanding of past climate conditions of our region.

Climate and Its Effects on Vegetation

The portion of the Cordilleran Ice Sheet that extended southward into what is now Washington, Idaho, and northwestern Montana constituted the Vashon glacier or Vashon Stade of the Fraser Glaciation. Part of the Vashon glacier known as the Puget Lobe was pushed southward between the Olympic Mountains and Cascade Range past what is now Olympia, Washington (Kruckeberg 1991, 287). The Puget Lobe covered the region of the Puget Lowland and reached its maximum extent approximately 15,000-13,500 years ago (Whitlock 1992, 5). The modern-day towns that lie on the border of the Puget Lobe in Washington are Eatonville, Vail, Rainier, Tenino, Little Rock, and Mima (Kruckeberg 1991, 287).

Before the glacier's full reach into this area, vegetation of the Puget Trough and adjacent lands was a mixture of grasses, pines (*Pinus*), firs

(*Abies*), American bistort-type (*Bistorta bistortoides*-type), sagebrush/ wormwood (*Artemisia*), and sedges (Cyperaceae), as well as herbs from the sunflower (Asteraceae) and carrot (Apiaceae) families (Tsukada and Sugita 1982, 404; Barnosky 1985, 266). This floristic assemblage is indicative of cold, dry conditions of tundra or subalpine nature (Barnosky 1985, 268; Tsukada and Sugita 1982, 405; Whitlock 1992, 6). The landscape would likely have been open and dry with existing taxa receiving more exposure to the elements than in a closed-canopy forest system. A contemporary comparative ecoregion to this historical reconstruction is possibly the northern Rocky Mountain subalpine zone (Barnosky et al. 1987 cited in Whitlock 1992, 6). During the time of the glacier's advance into the Puget Lowland, the climate became wetter and winter storms from the coast likely aided the extent of the glacier (Whitlock 1991, 9). The duration of ice cover over the Puget Lowland was a short 1,500 years (Kruckeberg 1991, 228).

A slight warming period following the full measure of glaciation contributed to the melting of the Puget Lobe and allowed for the glacier's recession northward. As this occurred, between 14,000 and 10,000 years ago, major geologic changes appeared on the landscape. For one, the rock and gravel that had been picked up by the glacier was left behind on top of the bare, cobbled substrate that was exposed by the scouring action of the glacier's advance. This deposit is known as glacial till or glacial outwash till. The northern drainage route at the Strait of Juan de Fuca was blocked by ice, so the meltwater found its outlet via the Chehalis River. This is evident today as the wide, but mostly unused, channel through which the Chehalis River flows; the channel was carved during this warming period to allow for the enormous volumes of meltwater to exit the lowlands to the Pacific Ocean (Kruckeberg 1991, 288-289).

At this time, a shift occurred in the fossil pollen record. New floral communities arrived in the areas that had been under the ice sheet, while those adjacent to the ice experienced changes in floral composition. Lodgepole pine (*Pinus contorta* var. *latifolia*) moved into de-glaciated areas, as it is adapted to life in poor soils (Whitlock 1992, 10). Later, Sitka spruce (*Picea sitchensis*), Douglas-fir (*Pseudotsuga menziesii* var. *menziesii*), western hemlock (*Tsuga heterophylla*), and mountain hemlock (*T. mertensiana*) moved in as well, indicating denser forest structures (Whitlock 1992, 10; Tsukada and Sugita 1982, 404). These forests were potentially broken up by clear, open areas as indicated by the *Artemisia* and grasses that still occurred in the pollen record at significant levels (Tsukada and Sugita 1982, 404; Barnosky 1985, 267). Mountain species may have kept a foothold in the wetter areas; however, the drier sites were likely taken over by taxa preferring milder temperatures and drier soils (Whitlock 1992, 11).

The early Holocene began approximately 10,000 years ago when the climate underwent a dramatic shift in what was the beginning of the Xerithermic period (xeri = dry). This climate change was caused by increased levels of summer solar radiation coupled with decreased levels of winter solar radiation due to the tilt of the earth on its axis (Whitlock 1992, 12). Temperatures during this period were roughly 4-5° C (7.2-9° F) warmer than they are today (Whitlock 1992, 18). Drought was characteristic of the time, as was a subsequent increase in fires (either in frequency or intensity) as indicated by greater amounts of charcoal found in the sediment layers (Cwynar 1987, 796; Whitlock 1992, 12).

The Xerithermic period is visible in the fossil pollen record as a sudden change in the presence or frequency of xeriphytic taxa, which then typically remained for the next 5,000 years. In the Puget Trough, increases in Douglas-fir and red alder (*Alnus rubra*) pollen, and bracken fern-type (*Pteridium*-type) spores occurred (Tsukada and Sugita 1982, 404; Barnosky 1985, 266; Whitlock 1992, 12), while there was a reduction of *Artemisia*, willow (*Salix*), lodgepole pine, spruce, mountain hemlock, and grass pollen. Multiple researchers attribute these changes either to an increase in temperature or drought conditions or both (Tsukada and Sugita 1982, 404; Barnosky 1985, 266). An important keystone species of today's Pacific Northwest prairie ecosystem, Oregon white oak (*Quercus garryana* var. *garryana*), began to move into the Puget Trough in greater numbers at this time (Tsukada and Sugita 1982, 404; Barnosky 1985, 266). Whitlock's climate reconstructions for the area near Battleground Lake (in the southern Puget Trough, beyond the Puget Lowland), "suggest that annual precipitation was 40-50% less than today between [9,500 years ago] and [4,500 years ago], and the annual temperature was 1-3° C [1.8-5.4° F] higher" (Whitlock 1992, 6). She also asserts that the path of migration for these xeriphytic species was from south to north.

As fire was prevalent at this time, adaptation to fire was an important component influencing survival as well as competition among species. While western hemlock and Sitka spruce are maladapted to fire, Douglas-fir and red alder quickly colonize even after intense burns (Whitlock 1992, 12). Bracken fern is also well adapted to fire and survives by storing energy in rhizomes protected underground. An increase in charcoal deposits in sediment layers (Tsukada and Sugita 1982, 407), coupled with the changing floral composition of this landscape, leads some scientists to believe that the landscape of these areas was probably patchy and relatively open, often referred to as "park-like". Cwynar (1987) believes it is likely that varying stages of succession existed simultaneously in a "mosaic", with fire being the limiting factor to successional advance towards mature forest of hemlock and spruce.

Sometime around 5,000 years ago, temperatures began to decrease while rainfall increased, corresponding to another shift in the fossil pollen record (Tsukada and Sugita 1982, 408; Barnosky 1985, 271; Cwynar 1987, 792; Whitlock 1992, 14). At this time during the late Holocene in both the northern and southern Puget Trough, western red cedar (*Thuja plicata*) began to grow. Western red cedar is adapted to humid or consistently wet conditions and does not fare well in dry, well-drained sites (Sudworth 1967, 157). Increases in western hemlock, western white pine (*Pinus monticola*), and Sitka spruce also occurred at this time (Whitlock 1992, 14). Oak began to decline at some sites (Tsukada and Sugita 1982, 404), while bracken fern (*Pteridium aquilinum* var. *pubescens*) began to taper off in others (Barnosky 1985, 266). Interestingly, at Mineral Lake, Washington, after a spike in charcoal fragments during the Xerithermic period, a decline occurred. However, this decline never consistently fell below average levels that existed before the Xerithermic, even though both the pre- and post-Xerithermic periods were relatively wet.

This changing charcoal abundance is possibly due to intentional burning practices by the Indigenous inhabitants of this area, who, according to anthropologists, arrived as early as 10,000 years ago (Storm and Shebitz 2006, 257). It is generally accepted that the extent of the Puget Trough prairies decreased around 5,000 years ago from their maximum extent during the Xerithermic period (Whitlock 1992, 14). The reliance of the Indigenous peoples on the ecology and structure of these prairies that had been formed over thousands of years was challenged by this period of increased moisture and subsequent shifts in floristic composition. There is an abundance of evidence indicating that Native peoples intentionally set fire to the prairies after the decline of environmentally-caused fires in order to maintain their open structure, biotic communities, and prevent the encroachment of the surrounding forests (Norton 1979, 177-189; Leopold and Boyd 1999, 145-154; Devine and Harrington 2006, 63-65; Hosten et al. 2006, 64; Storm and Shebitz 2006, 257).

Current Climate

Today in the Pacific Northwest, we experience the ecological effects of the climate shifts that occurred during the late Holocene. Notwithstanding immediate climate change due to human-caused increases in atmospheric carbon, this period has been one of mild, dry summers and cool, wet winters characteristic of a Mediterranean environment. From October to March, high pressure air masses from the Pacific Ocean near the Gulf of Alaska meet low pressure air masses from off the coast of southern California and deliver high-humidity air to the Olympic Mountains and Cascade Range with prevailing winds from the south or southwest (Western Regional Climate Center 2014).

These air masses rise as they meet the mountains and cool in temperature, dropping their moisture in the form of rain and snow. This climate supports the dominance of coniferous tree species (Tsukada and Sugita 1982, 401), potentially attributed to the conifers' ability to withstand long periods with very little rainfall (i.e., Pacific Northwest summer drought) (Kuchler 1946, 138-141).

West of the Cascades, the Puget Trough lies in the rain shadow of the Olympic Mountains and therefore receives significantly less rainfall than the mountainous areas directly to the east and west. The lowest areas of the Puget Trough have a maritime climate with dry summers and high rainfall throughout the winter months, during which temperatures generally remain above freezing, allowing for little snow (Puget Sound Institute 2014). Overall average annual precipitation from 1931-2005 taken from three weather stations (Olympia, Centralia, and Oakville) near the south Puget Sound prairies was 51.57 inches (131 cm) (Western Regional Climate Center 2014), with the vast majority falling between September and April (Constance et al. 2003, 3). Annual average high and low temperatures from the same three weather stations and the same time period were 61.2° F (16.2° C) and 40.6° F (4.8° C), respectively (Western Regional Climate Center 2014). Months with the highest temperatures are typically July and August with lows falling in December and January.

Future Climate

Historical climate reconstructions based on fossil pollen records are potential clues to the floristic communities of a future affected by anthropogenic climate change. Although today there are many factors influencing floristic communities on a landscape scale that were nonexistent before European and Euro-American settlers arrived to the Puget Sound Basin, our local remnant prairie-oak mosaics may contain some of the species that will colonize our ever-changing environment. The study of historical floristic composition allows us to view changes over vast temporal scales and gives us an opportunity to infer how future changes in temperature and precipitation may affect current ecological communities. Although many species have come and gone from the Puget Sound ecoregion since the advent of European and Euro-American settlement, we can now attempt to make intelligent land-use and management decisions that support biodiversity and the potential for healthy local ecosystems.

Geology
Patrick Pringle

Introduction and Physiography

The Puget Lowland is a sedimentary basin complex that is bounded on the west by the Olympic Mountains and Black Hills and on the east by the Cascade Range. The complicated geology of the basin records tectonic deformation, changing sedimentary environments, and active geomorphic processes. Tectonic deformation has occurred and continues with the movement of continental and oceanic plates. These tectonic plates form the lithosphere, which consists of the Earth's crust and the uppermost part of the mantle. For a brief explanation of plate tectonics and other technical terms in this section, please see the Glossary.

Geologic deposits in the region record both the multiple advances of continental glaciers over the Pleistocene (~2.6 million—11,500 years ago), glacial outburst floods, volcanism of Cascade Range volcanoes, and landscape processes of the interglacial periods. The southern Puget Lowland encompasses most of the tectonic basin that extends roughly from the Tacoma Fault at about the latitude of Tacoma to the higher elevation prairies located slightly south of Chehalis.

Geology and Geologic History of the Puget Lowland

The Georgia Basin and Puget-Willamette Lowlands are portions of an active tectonic basin in the crust of the North American tectonic plate above the subducting Juan de Fuca plate. These elongated, sedimentary basins, known as forearc basins, occur between the resulting subduction-related trench and the volcanic arc known as the Cascade Range.

The structural controls on the Georgia Basin and Puget-Willamette Lowlands are complicated because subduction is accompanied by a northerly compression caused by the coupling of the North American plate with the Pacific plate, which is moving to the north-northwest and creating zones of northwest-trending megashear (Barnett et al. 2010). This megashear combined with subduction results in tectonic compression to the north or northwest in the Puget Lowland. Velocities of the crust derived from global positioning system (GPS) measurements show a northward or northeast motion of as much as one centimeter per year in some locations in the southern Puget Lowland and higher rates toward the coast. Thus the shallow crust above the subducting slab is being slowly shortened in a northerly direction in the lowland, and in response, structures such as the east-west-trending

Figure 2. Map of the southern Puget Lowland showing major physiographic boundaries and fault and fold structures. Dashed lines show inferred faults. Geology shown is from Washington Division of Geology and Earth Resources Geoportal 1:250,000-scale geology layer and Walsh et al. (1987). Derived from public domain material from the Washington Department of Natural Resources.

Seattle and Tacoma fault systems have episodically ruptured in response to the strain buildup. A number of faults undoubtedly lie undiscovered.

The shallow crustal faults are responsible for both uplift and subsidence that have segmented the basin. The southern Puget Lowland prairies are largely within the Tacoma basin, whose southern margin is not well constrained, but could be delineated by the Bald Hills, the prairie upland slightly south of Chehalis, and portions of the Willapa Hills. Highlands that form the southern margin of the southern Puget Lowland are likely also controlled at least somewhat by faults and folds, but detailed geologic mapping in this area is lacking (Fig. 2). The northern boundary of the basin is the Tacoma Fault (Gomberg et al. 2010; Sherrod 2001).

The southern Puget Lowland has been glaciated at least seven, and perhaps as many as twelve times during the Pleistocene Epoch (Easterbrook 1994a; Easterbrook 1994b; Troost and Booth 2008; Kathy Troost pers. comm. 2013). The Puget Lobe of the Cordilleran Ice Sheet is the most recent major glacial advance, taking place during the Vashon Stade of the Fraser Glaciation. The Puget Lobe of the Vashon glacier flowed south between the Olympic Mountains and Cascade Range advancing and retreating as it moved (Haugerud et al. 2003). The Puget Lobe reached at least as far south as Rochester-Maytown in the southern Puget Lowland 16,900–16,500

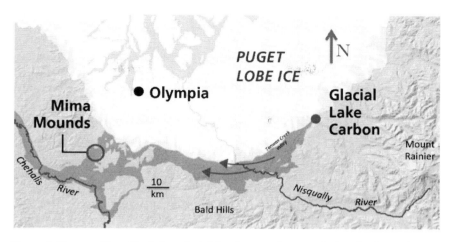

Figure 3. Tanwax-Ohop Creek late-glacial flood materials entered Glacial Lake Carbon via Nisqually River and were moved westward to Mima Mounds and the Chehalis River. The relatively high percentage of andesite in the deposits of the late-glacial flood distinguishes them from outwash originating from meltwater streams under and within the Puget Lobe of the Vashon Glacier, which contain lesser andesite and a higher percentage of metamorphic rocks. Derived from public domain material from the Washington Department of Natural Resources. Arrows show flow path of the late-glacial flood.

years ago (Porter and Swanson 1998; Walsh et al. 2003) and stayed near its maximum extent for only a few centuries, retreating to north of Seattle by as early as 16,400 years ago (Troost and Booth 2008). Booth (1994), Booth and Goldstein (1994), and Booth et al. (2003) offer important insights about the role of the glacier and glacial meltwater in the creation of the southern Puget Lowland landscape.

Geologic evidence shows that enormous lakes that were impounded by the Puget Lobe episodically breached and flooded areas of, and near the ice margin. These outburst floods, or jökulhlaups, eroded channels along the ice margin and across the glaciated landscape and deposited large amounts of gravelly materials. Jökulhlaup is an Icelandic word for a glacial outburst flood, which could be pronounced "yo-kul-hloyp". One of the earliest great late-glacial outburst floods originated at glacial lake Carbon, which occupied the valley of the Carbon River and was impounded near Carbonado; the Ohop-Tanwax Creek valley flood and debris flow resulted from breaching of lake Carbon (Fig. 3). The flood drained through Fox Creek in the Cascade Range (east of the hamlet of Electron) and rushed southeasterly across the ice margin and around the toe of the Puget Lobe, probably triggering a large earthflow, first noted by Crandell (1963), that liquefied and flowed as a debris flow to the southwest following channels such as the Tanwax Creek valley that had been carved by the flood (Pringle et al. 2001; Pringle and Goldstein

2002; Goldstein et al. 2010).

Surficial Materials of the Southern Puget Lowland Prairies

Drift deposits of the Vashon glaciation record landscape changes that occurred as the ice margin advanced and retreated. The word "drift" is commonly used to refer to any glacial deposit regardless of how it was deposited. These deposits include lake sediments (clay silts), advance and recessional outwash material (stratified sands and gravels), and till, the poorly-sorted mixture of sediment from under, within, and/or atop the ice. Rarely is the entire sequence of these deposits found in one exposure. If it were, the sequence from bottom to top would be lake sediments, advance outwash, till, recessional outwash, and lastly lake sediments.

In late-glacial and postglacial times, lake and stream sediments were deposited atop the glacial drift. During its maximum extent, the ice of the Puget Lobe dammed streams along the ice margin, forming large lakes. For example, Bretz (1913) first noted clayey lake sediments that recorded large lakes that had formed along the retreating margin of the Puget Lobe. In the Thurston County area, deposits of glacial Lake Bretz are found at elevations as high as ~43 m. Breaching of glacial Lake Carbon, noted above, released a flood of water and sediment that incised Ohop and Tanwax Creeks along the margin of the Puget Lobe (Pringle and Goldstein 2002; Goldstein et al. 2010). The flood cut deep channels in and through the glacial drift locally and left deposits rich in the volcanic rock called andesite and other sediments of Cascade origin as far downstream as Rochester and Centralia. Such floods were responsible for the abundant andesite rocks in locations such as Rocky Prairie, Bucoda, and in the Chehalis River valley near, and north of, Centralia.

Deposits of the Olympia nonglacial (interglacial) interval (65,000–17,000 years ago) commonly underlie the Puget Lobe drift. The typical nonglacial deposits include peats and alluvial silts, sands, and gravels. During nonglacial times, rivers in the southern Puget Lowland carried sediments from both the Mount Rainier area and from the Cascade Range east of Centralia, the Olympic Mountains, and the Willapa Hills. But by far Mount Rainier and the Cascade Range near it were the dominant sources of sediment—thus it is not uncommon to find sediments that are rich in andesite fragments (typically red, gray, and black lithic grains) and the crystals of minerals typically associated with andesite, such as plagioclase feldspar. In addition to offering criteria for differentiating glacial from nonglacial deposits, Borden and Troost (2001) further noted that the percentage of land area receiving deposition during nonglacial periods was small relative to that of the glacial periods.

Puget Lobe glacial sediments are dominated by rocks and minerals that have sources that are typically rich in quartz and mica from the Canadian cordillera (chain of mountain ranges) and central and north Cascade Range.

However, along the west side of the southern Puget Lowland, these deposits can also contain some volcanic rock in the form of basalt from the Olympic Mountains and Black Hills. Likewise, glacial sediments from nearby Mount Rainier also contain andesite and other rocks of Cascade Range origin.

The outburst floods from glacially-dammed lakes appear to be the origin of deposits in many of the outwash plains that now host prairie landscapes (Troost 2007). In recent detailed mapping of the East Olympia (Logan and Walsh 2005) and Maytown (Logan et al. 2009) Quadrangles, the authors used lidar to distinguish the surficial geology in great detail at 1:24,000 scale. Lidar can map elevations with precision on the order of one decimeter, allowing delineation of terrace features and landforms (and deposits) of distinct ages and origins as well as post-glacial deposition or landscape adjustments owing to mass movement or faulting (Haugerud et al. 2003). Thus, detailed mapping of the surficial geologic materials via site investigations, lithologic evaluation of sediment, and lidar mapping should yield an improved understanding of the characteristics of taxonomically classified soils such as those delineated by Pringle (1990).

Mima Mounds

It would be negligent to discuss the geology of the surficial geologic materials of the southern Puget Lowland prairies without mentioning the enigmatic Mima Mounds. A number of investigators have speculated about their formation: Washburn's 1988 treatise gives a highly informative overview of the Mima Mounds, their characteristics, and their distribution, and also reviews a number of hypotheses about their formation. In his pioneering publication, Bretz (1913) suggested the mounds could owe their origin to deposition of sediment into suncups formed in glacial snow and ice. Berg (1989, 1990, 1991) speculated on a possible earthquake trigger for mound formation, while Burnham and Johnson (2012) propose a biological origin for the mounds, namely giant, mound-forming rodents. Traditional stories involving supernatural beings also exist to explain their creation, including ones that involve Paul Bunyan, while others point to Native women dumping piles of rocks under Hummingbird's direction. For geologists, the details of their genesis remain a mystery and a focus of ongoing research.

Pringle and Goldstein (2002) and Goldstein et al. (2010) noted the poorly-sorted deposits (diamicton textural character) of the Mima Mounds and the high percentage of andesitic material in them and suggested that the geologic material in the mounds likely was deposited by the Tanwax-Ohop Creek late-glacial flood and debris flow event noted above (Figs. 4 and 5). Walsh and Logan (2005, Fig. 1) and Logan et al. (2009, Fig. 3) assigned the Tanwax deposits to a Quaternary unconsolidated deposit designated as Qgoy3, which is an abbreviation for Quaternary glacial outwash, with y3 being a reference to its youth relative to other recessional outwash deposits from the Pleistocene. In their mapping, they offer this detailed description of unit Qgoy3:

> Soils capping bedrock areas tend to be reddish brown, but those capping most glacial outwash deposits tend to be very dark brown to black. The soils covering unit Qgoy3 are formed into mounds about 2 to 6 ft high and 10 to 30 ft across. The mounds are referred to as Mima mounds and have been extensively studied by previous workers (summarized in Washburn, 1988; and discussed in Bretz, 1913). It is notable that unit Qgoy3 is the only unit within the East Olympia quadrangle on which the Mima mounds formed (this study; Pringle and Goldstein, 2002). The mounds must have formed very shortly after unit Qgoy3 was deposited, because (1) they did not form within kettles, which clearly formed after unit Qgoy3 was deposited (kettles crosscut melt water channels); (2) they appear to be partially buried by alluvial fans near the margins of the unit Qgoy3 channel; and (3) they did not form in unit Qgoy4 channels that cut the surface of unit Qgoy3.

The significance of recognizing the high percentage of andesite cobbles and massive, matrix-supported, debris-flow-like texture of most Mima mounds is that deposition of the sediment, particularly that of the mounds comprising unit Qgoy3, can be tied to the breaching of a late glacial ice-marginal lake and the resultant Tanwax flood events, and thus fixed in time. Further analysis of the mounds, including those on different (younger) terraces that might be associated with post-Tanwax-Ohop Creek glacial outburst floods, will probably reveal more clues as to how the mounds formed.

Figure 4. An excavated mound at Mima Prairie showing diamicton-like texture of mound material (darker and carbon-rich) and coarser texture of lower cobble gravel, which locally displays cross bedding and horizontal bedding (indicative of water-laid deposit).

Figure 5. Close up of a poorly sorted deposit comprising a mima mound in Rocky Prairie showing abundant rocks composed of broken pieces of older andesite rocks and the matrix supporting these clasts. Trenching tool is 45 cm long.

Where to Find Information on Surficial Materials in the Puget Lowland

For reconnaissance purposes, probably the best place to find geologic information is through the Washington Division of Geology and Earth Resources (DGER) web pages. Geologic maps at 1:500,000 scale are accessible via the Washington State Geologic Information Portal, which has a GIS-type interface. Smaller scale maps (1:250,000) are available in the Google Earth kmz files of geology published by county (via the DGER GIS Data link). The Index Map of geologic mapping accesses maps of different scales. The 1:100,000-scale maps are simply scans of the original black and white maps. The most up-to-date maps, with most detail and the status of geologic mapping at 1:24,000-scale, can be found via the Index to Geologic and Geophysical Mapping of Washington. For a geologic map that covers the area under discussion, see Walsh et al. (1999). For more general descriptions of the geology of the Puget Prairies, see Kruckeberg (1991 and 2002). Figures 1 and 3 of Walsh and Logan (2005) and Logan et al. (2009) respectively (available online) show detailed interpretations of outwash channels of various ages/units based on the lidar, and Fig. 4 in Logan et al. (2009) shows mapped locations of the Mima Mounds on a low-resolution relief map. The following are reliable internet sources of information on landscape, geology, and soils:

WADNR Division of Geology
  www.dnr.wa.gov/geology

Index map for accessing geologic maps of Washington at various scales
  www.dnr.wa.gov/programs-and-services/geology/geologic-maps/surface-geology#get-our-maps

WADNR Division of Geology 3-D geology and surface overlays
  www.dnr.wa.gov/programs-and-services/geology/geologic-maps/3d-geology

Soil maps
  websoilsurvey.nrcs.usda.gov/app/HomePage.htm

Washington soils information and maps
  www.nrcs.usda.gov/wps/portal/nrcs/main/wa/soils/

Crude map of southern Puget Lowland prairies
  www.nature.org/idc/groups/webcontent/@web/@washington/documents/media/prd_010752.jpg

CNLM
  www.southsoundprairies.org/
  www.southsoundprairies.org/visit-the-prairies/

Soil extent mapping tool
  apps.cei.psu.edu/soiltool/

High-resolution lidar topography of the Puget Lowland, WA
  earthweb.ess.washington.edu/~bsherrod/brian/Haugerud_etal_GSA-Today_LiDAR_Bonanza.pdf

Note: Landform or geologic unit names have been capitalized in conformance with usage in the U.S. Geological Survey (USGS) Geographic Names Information System, with the USGS National Geologic Map Database, and with previous usage in Washington State geologic publications.

Vegetation

Joe Bettis

The lowland prairies of the south Puget Sound region are among the most interesting landscapes found in western Washington. These prairies are distinctive both floristically and structurally from the coniferous forests that dominate the region. As with the development of other kinds of plant communities, contemporary local prairies are a product of biotic and abiotic factors playing out over millennia (Kruckeberg 2002). In the south Puget Sound region, four influences in particular shape modern prairie vegetation: the presence of glacial outwash deposits, past climate change, cultural burning practices of Native Americans, and changes in the landscape initiated by the arrival of Euro-American settlers.

Continental glaciation approximately 15,000 years ago left deposits of glacial outwash at the shifting southern terminus of the Vashon lobe of the Cordilleran Ice Sheet (Kruckeberg 1991). Prairie sites occupy these flat to mounded deposits of glacial outwash upon which fairly coarse and well-drained soils have formed (Franklin and Dyrness 1973). These outwash-derived, gravelly to sandy soils are classified as the Spanaway and Nisqually series (Lang 1961). For more information on soils, county soil surveys are available through the Natural Resources Conservation Service's website (NRCS n.d.). These soils are home to threatened subspecies of the Mazama pocket gopher (*Thomomys mazama*), and the mapping of these and related soils is being used to inform conservation decisions regarding prairie-dependent species (USFWS 2014). Unique patterned landscapes of Mima mound formations, the origin of which has puzzled researchers for more than a century, occur on glacial outwash deposits (Kruckeberg 1991). Recent studies suggest that an interaction of vegetation and soils along with erosion and sediment deposition processes may be behind the formation of the mysterious mounds (Cramer and Barger 2014).

Analysis of ancient pollen accumulations from lakes and bogs has allowed scientists to reconstruct the historical vegetation of the south Puget Sound, revealing clues about past climate change in the Pacific Northwest (Hansen 1947). By relating the presence of pollen from particular plant species with known environmental tolerances, historical climatic shifts may be inferred. Ages of pollen strata are assigned, in part, by layers of volcanic ash found within the lakes' sediment profiles from known eruptions (Whitlock

1992). Following the glacial maximum (20,000 to 14,000 years ago), local lowland vegetation was composed of those species adapted to cool and dry conditions, such as those found today in grassland and shrub-steppe habitats east of the Cascade crest (Whitlock 1992). The period from 14,000 to 10,000 years ago saw a rise in temperate plant species, suggesting more precipitation and warmer temperatures (Whitlock 1992). South Puget Sound prairie vegetation is thought to have developed during a period of warming and increased aridity known as the Hypsithermal or Xerithermic, between 5,000-10,000 years ago. During this period the pollen record shows increases in modern prairie plants such as oak and camas (Whitlock 1992).

Fires set by Native Americans to promote the production of food plants and open hunting grounds have influenced the historical persistence of prairie vegetation. The development of a cooler, wetter climate about 4,500 years ago is thought to have been the initial impetus for Indigenous burning practices (Leopold and Boyd 1999). This transition period, which ushered in our current climate regime, saw an increase in the development of coniferous forests (Barnosky 1985). Seasonal burning by Native peoples thwarted the encroachment of trees into the prairies where significant food products were tended (Leopold and Boyd 1999). In quantitative field experiments, burning at one to two year intervals has been shown to promote the recruitment of camas, an important native food plant (Storm and Shebitz 2006).

The arrival of Euro-American settlers nearly two centuries ago impacted prairie vegetation almost immediately (Barnosky 1985). In the absence of regular, low-intensity fires, many prairies have succumbed to invasions of Douglas-fir and other trees and shrubs. The intentional and unintentional introduction of weedy species by settlers has also had a major impact on the integrity of native prairie ecosystems (del Moral and Deardorff 1976). Field studies have illustrated that increased abundance of non-native grasses, in particular, has a negative effect on the richness of native prairie species present at sites where frequent fire disturbance is absent (Hill and Fischer 2014).

Glacial Outwash Prairie Vegetation

High quality prairies of the south Puget Sound are dominated by the native bunchgrass Roemer's fescue (*Festuca roemeri*) (Chappell 2006). Other native bunchgrass species such as California oatgrass (*Danthonia californica*) and pine bluegrass (*Poa secunda*) may also be present. Among the tussocks of bunchgrass a spectacular wildflower display occurs annually during May and June. At this time, broad, blue swaths of common camas (*Camassia quamash*) contrast with the golden flowers of western buttercup (*Ranunculus occidentalis* var. *o.*) and deltoid balsamroot (*Balsamorhiza deltoidea*). Other native wildflowers likely to be encountered include Oregon sunshine (*Eri-

ophyllum lanatum var. *leucophyllum*), spring gold (*Lomatium utriculatum*), sickle-keeled lupine (*Lupinus albicaulis*) and early-blue violet (*Viola adunca* ssp. *adunca*). Small patches of Oregon white oak (*Quercus garryana* var. *garryana*) and Douglas-fir intersperse the otherwise treeless landscape.

Non-native plant species of Eurasian origin are quite common in more disturbed prairies, with forbs such as hairy cat's-ear (*Hypochaeris radicata*), oxeye daisy (*Leucanthemum vulgare*), St. John's wort (*Hypericum perforatum*), and sheep sorrel (*Rumex acetosella*) broadly distributed among extant prairie sites. Exotic pasture grasses including tall oatgrass (*Arrhenatherum elatius*), sweet vernalgrass (*Anthoxanthum odoratum*), common velvet grass (*Holcus lanatus*), and colonial bentgrass (*Agrostis capillaris*) are often dominant, competing with native grasses and forbs alike (Chappell 2006).

An introduced shrub, Scotch broom (*Cytisus scoparius*), is a regular invader of the prairie ecosystem, posing a serious threat to native vegetation by modifying soil chemistry, making sites less hospitable to native plants (Haubensak and Parker 2004). Adapted to disturbed and mineral soils, Scotch broom, like other legumes, hosts nitrogen-fixing bacteria in its roots that are able to convert atmospheric nitrogen (N_2) into a form of nitrogen accessible to plants (NH_{4+}) (Raven, Evert and Eichhorn 2005). Native prairie plants are adapted to nitrogen-poor soils, and the addition of even small amounts of nutrients can promote the recruitment of non-native vegetation (Haubensak and Parker 2004). Field studies indicate that the addition of sucrose to plots impacted by Scotch broom dominance may promote native species by reducing plant-available nitrogen, subsequently offering a temporary reduction in the cover of non-native species (Kirkpatrick and Lubetkin 2011).

Prairie grasslands in the south Puget Sound are one of Washington's most endangered habitats, currently occupying only a small fraction of their former range. A model of the historical extent of prairies has been established by mapping the presence of soils that have developed under prairie vegetation (Crawford and Hall 1997). Using this method, the Washington Natural Heritage Program found that of the 149,360 acres featuring prairie soils, only 12,582 acres currently supported prairie plant communities. Of these remaining prairies only 2,993 acres were dominated by native plants, indicating that the majority of extant prairie sites are highly degraded (Crawford and Hall 1997).

A compilation of species lists for south Puget Sound prairies indicates 278 species of vascular plants are known to inhabit south Puget Sound prairies, of which 59 percent are native taxa (Dunwiddie et al. 2006). Though the complete historical flora of south Puget Sound prairies is unknown, researchers have compared modern accounts of our local prairie flora to historical records and herbarium data to reveal the extent of change in the composition of prairie vegetation. Native annual plants, an important functional group, have largely disappeared when compared to their historical presence on south Puget Sound prairies (Dunwiddie et al. 2014). Several sensitive and range-restricted species rely on prairie habitat, including golden paintbrush (*Castilleja levisecta*), small-flowered trillium (*Trillium parviflorum*), Mardon skipper butterfly (*Polites mardon*), Taylor's checkerspot butterfly (*Euphydryas editha* ssp. *taylori*), Oregon spotted frog (*Rana pretiosa*), streaked horned lark (*Eremophila alpestris strigata*) and local subspecies of the Mazama pocket gopher (NatureServe 2015). Conservation and ecological restoration efforts in the south Puget Sound prairies on publicly and privately managed preserves include invasive species control measures, propagation and reintroduction of native species, and prescribed burns (Sinclair et al. 2006).

Wet Prairies

Less extensive than those found in Oregon's Willamette Valley, wet prairies (sometimes referred to locally as wet swales) are fairly limited in the south Puget Sound. Typical prairie soils of the south Puget Sound contain a coarse matrix including a high percentage of sand and gravel, and are quite quick-draining (Lang 1961), in contrast to the often clay-rich, poorly drained soils found in Willamette Valley wet prairies. Wetter sites in these glacial outwash soils are limited to low areas along drainage courses where the ground water is closer to the surface, or localities featuring a soil layer or substrate that retards drainage, allowing water to remain near the surface (Easterly et al. 2005). The Scatter Creek Wildlife Area located 20 miles south of Olympia features a seasonally wet prairie that hosts an interesting assemblage of plants including narrow-leaved mule's ears (*Wyethia angustifolia*), American bistort (*Bistorta bistortoides*), northern bedstraw (*Galium boreale*), tufted hairgrass (*Deschampsia cespitosa*), and several species of sedges (*Carex* spp.). These wet prairie sites are thought to have once been more widespread, especially in the upper Chehalis River valley (Storm 2004).

Oak Savannas

In the south Puget Sound, open and sparsely treed habitats dominated by herbaceous plant species are quite rare. Historically, oak and conifer savannas may have been more common, maintained by the burning practices of Native Americans (Chappell and Crawford 1997). Although rare, some high quality savannas persist in areas that may initially seem unlikely, such as those located at Joint Base Lewis-McChord (JBLM) in southwestern Pierce County. Fires set during the course of military training exercises inhibit the recruitment of woody species, although mature trees can survive these low-intensity ground fires. JBLM also supports a savanna of ponderosa pine (*Pinus ponderosa* var. *ponderosa*), a species that is rare west of the Cascade Range in Washington. Typically occupying glacial outwash deposits, many of these and other lightly-wooded sites were likely prairie at one point and may be viewed as intergradient between open habitat and forest (Chappell and Crawford 1997; Chappell 2006).

The composition of herbaceous species in south Puget Sound savannas is similar to that of prairie habitats, with the addition or increase in shade tolerant species such as long-stolon sedge (*Carex inops* ssp. *inops*) and blue wildrye (*Elymus glaucus* ssp. *glaucus*). With the cessation of fire since Euro-American settlement, savanna habitats trend toward oak woodlands or forests of Douglas-fir (Chappell 2006).

Oak Woodlands

Often adjacent to prairies are hardwood-tree-dominated zones known as woodlands. The canopy of a woodland is fairly open above a moderately well-developed mix of forbs and shrubs. In the lowlands of south Puget Sound, common woodland trees include Oregon white oak and Douglas-fir. Oregon ash (*Fraxinus latifolia*) can also be an important component of moist oak and riparian woodlands in this area (Chappell and Crawford 1997). Historically, oak woodlands were more broadly distributed (Devine and Harrington 2006), persisting under a regime of low intensity ground fires set by Native peoples. Woodland stands of oak likely develop when the oaks invade open prairie sites (Chappell 2006). Soon after oak invasion of the prairies, herbaceous plant composition remains similar to that of the open prairie, with common camas, yarrow (*Achillea millefolium*), and Henderson's shooting star (*Dodecatheon hendersonii*) present. Shrub layers often feature common snowberry (*Symphoricarpos albus* var. *albus*) and serviceberry (*Amelanchier alnifolia*), in addition to hairy honeysuckle (*Lonicera hispidula*) in south-facing oak woodlands. Without fire, these sites develop a more extensive shrub layer and eventually lead to the establishment of Douglas-fir (Chappell 2006). Douglas-fir will overtop and stunt the growth of the shade-intolerant oaks (Franklin and Dyrness 1973). Research on the restoration of oak habitat suggests that shade-stunted oaks respond positively once released from the shade of Douglas-fir (Devine and Harrington 2006). Oak woodlands are an important element of western gray squirrel (*Sciurus griseus*) habitat, a Washington state threatened species (NatureServe 2015).

Visiting the Prairies

The unique vegetation of south Puget Sound prairie ecosystems is accessible to the public at several local preserves. Just west of Littlerock, Washington, Mima Mounds Natural Area Preserve (NAP) offers trails that tour a mounded prairie landscape, as well as mixed oak and Douglas-fir forest. Mima Mounds NAP, managed by Washington Department of Natural Resources, is home to a wide diversity of plants and wildlife and has been an active research site for over 50 years. Scatter Creek Wildlife Area, managed by the Washington Department of Fish and Wildlife, accesses upland and seasonally wet prairies as well as a variety of forested habitats. Scatter Creek Wildlife Area is located just north of Grand Mound, Washington. Those interested in visiting these or other prairie sites in the south Puget Sound are encouraged to contact the particular site's managing entity to determine the best time to visit and whether any restrictions may be in place. Most prairie preserves host volunteer work events, at which the public can take an active role in the restoration of these unique ecosystems.

Sensitive Species

Lisa Hintz

Vast and sudden changes to the Pacific Northwest landscape came as a direct result of European and Euro-American settlement and land-use practices. These changes quickly affected the south Puget Sound prairie-oak landscapes in many ways. Broad, persistent effects include reduction of the overall expanse of prairie ecological communities and the loss of corridors connecting those communities together. Due to the relative rarity of the prairies in a region dominated by coniferous forest, a variety of plants and animals associated with this ecosystem are currently at risk.

Even before the advent of the Endangered Species Act in 1973, states had compiled lists of species found within their borders using available distribution and abundance information in order to determine their relative rarity. Since 1977, the Washington Natural Heritage Program (WNHP), within the Department of Natural Resources, has evaluated Washington State's plant list every two years, determining the status of rare species based on abundance, vulnerability, threats, and other criteria in relation to their risk of extirpation from the state or extinction (Camp and Gamon 2011, 4-5). The WNHP is part of the NatureServe network, a non-profit organization for all fifty states' Natural Heritage Programs as well as similar programs in all Canadian provinces and some Caribbean and Latin American nations. NatureServe determines global ranks and WNHP assigns sub-national ranks characterizing rarity and endangerment for these species. Global and state ranks are denoted using a number system from 1 (critically imperiled) to 5 (demonstrably secure). The United States Fish and Wildlife Service also assigns status to species based on rarity, using information from a variety of sources. All of these determinations inform legal and practical approaches to conservation concerning environmental planning and management. See Table 1 for a partial list of the rare plant and animal species found on the south Puget Sound prairies and their conservation designations as of January 2015 (Arnett 2015). For a good introduction specific to the rare plants of Washington, see Camp and Gamon (2011), particularly pages 4-9. The current global, state, and federal ranks for sensitive species can be found at these websites:

www.explorer.natureserve.org/servlet/NatureServe?init=Species
www1.dnr.wa.gov/nhp/refdesk/lists/plantrnk.html

Sensitive Species

	Common Name	Scientific Name	NatureServe State Rank	NatureServe Global Rank	Federal Status	WNHP State Status
PLANTS	golden paintbrush	*Castilleja levisecta*	1 critically imperiled	1 critically imperiled	threatened	threatened
	white-topped aster	*Sericocarpus rigidus*	3 very rare and range restricted	3 very rare and range restricted	species of concern	sensitive
	Hall's aster	*Symphyotrichum hallii*	1 critically imperiled	4 apparently secure	n/a	threatened
ANIMALS	Taylor's checkerspot butterfly	*Euphydryas editha taylori*	1 critically imperiled	1 critically imperiled	endangered	endangered
	streaked horned lark	*Eremophila alpestris strigata*	1 critically imperiled (breeding status)	2 imperiled	threatened	endangered
	Mardon skipper butterfly	*Polites mardon*	1 critically imperiled	2/3 imperiled/very rare and range restricted	species of concern	endangered
	Mazama pocket gopher	*Thomomys mazama*	2 imperiled	4 apparently secure	threatened	threatened

Table 1. Partial list of the rare plant and animal species found on the south Puget Sound prairies. Note that numerous other plant taxa with rare status occur in the prairies.

VASCULAR PLANTS OF THE SOUTH SOUND PRAIRIES

The following is a brief introduction to three plants, two butterflies, one bird, and one mammal associated with the south Puget Sound prairies. This list is by no means complete; there are many other rare and at-risk species found on the south Puget Sound prairies. For more extensive information, refer to Camp and Gamon (2011), Stinson (2005), and the Washington Department of Fish and Wildlife's annual report entitled "Threatened and Endangered Wildlife in Washington."

Plants

Golden Paintbrush:

Golden paintbrush (*Castilleja levisecta*) was federally listed as threatened in 1997 and still maintains that status (Caplow 2004, 3). At the state level, it is threatened. This colorful perennial herb is endemic to Puget Trough prairies from southern British Columbia to central Oregon. Historical herbarium records indicate its former range was limited to the prairies of Washington, Oregon, and British Columbia (Lawrence and Kaye 2006, 140; Consortium of Pacific Northwest Herbaria 2014). It is short lived (5-6 years) and known to reproduce exclusively by seed. Golden paintbrush has roots that parasitize neighboring plants (Caplow 2004, 3; Kaye and Blakeley-Smith 2008, 1), but since it also contains chlorophyll and photosynthesizes, it is not entirely reliant on a host plant to survive (a trait known as hemiparasitism). However, studies have shown it to be more productive with a host, and different host species may have varying effects on individuals (Caplow 2004, 6). Golden paintbrush is known to parasitize Roemer's fescue, red fescue (*Festuca rubra*), and Oregon sunshine (*Eriophyllum lanatum* var. *leucophyllum*) (University of Washington 2006) as well as tufted hairgrass (*Deschampsia caespitosa*)(Hamman et al. 2015). These ecological relationships are important because the level of productivity of golden paintbrush may be more greatly determined by associations with neighboring plants than that of species that do not have parasitic relationships.

Plants rely on pollination and seed dispersal for their success. However, neither are well understood for golden paintbrush. Based on field observations, a species of bumblebee (*Bombus californicus*) is a likely pollinator for golden paintbrush and potentially the only pollinator (Caplow 2004, 8 and 25-26). This bumblebee uses abandoned rodent tunnels for nesting in areas with grass thatch; thus managing for golden paintbrush also means the habitat of this bee species must be taken into consideration, especially if it is indeed the sole or primary pollinator (Caplow 2004, 25-26).

Golden paintbrush is nearly incapable of self-fertilization, and research shows that individuals have higher numbers of fully developed seeds if they are fertilized by plants from a separate population (Caplow 2004, 8). In order to support greater seed viability in the wild, populations of golden paintbrush should be close enough together and have functional corridors within the flight ranges of pollinators. Better understanding of the life cycle and ecological interactions of the associated bumblebee may assist restoration efforts of golden paintbrush. Furthermore, it is presumed that seeds of golden paintbrush are transported only a short distance from the parent plant, dispersed when knocked out of the fruits by passing animals or high winds (Caplow 2004, 9).

Threats to golden paintbrush are typically those that threaten the prairies overall: urban and rural development, agriculture (including some forms of livestock grazing), as well as the short- and long-term effects of fire suppression. Multiple studies have shown that golden paintbrush responds well to burning (Dunwiddie et al. 2001 cited by Caplow 2004, 26-27). However, outplanting of golden paintbrush plugs (small, potted plants) has been the primary method for reintroducing golden paintbrush to historical sites (Dunwiddie 2009, 3). More recently, as volunteers and managers have collected wild seed and increased seed production, seeding has become the primary strategy for golden paintbrush recovery. Recent outplanting and seeding efforts at Glacial Heritage, Tenalquot, and Mima Mounds have contributed to the growing knowledge of how to establish self-sustaining, genetically robust populations of this species, which is now on a path toward delisting within the next five years.

White-Topped Aster:

White-topped aster (*Sericocarpus rigidus*) is found in several south Puget Sound prairies, with the majority of its populations at Fort Lewis (Giblin 1997, 93). It is state ranked as sensitive by WNHP, and federally it is a species of concern. Although some native prairie taxa are found in other parts of North America, white-topped aster is endemic to the remnant low elevation prairies ranging from Vancouver Island south to the Willamette Valley west of the Cascades (Thomas and Carey 1996, 152). While other species in this genus are generally found in the forest understory, white-topped aster is limited to grasslands (Giblin 1997, 94). Stands can be substantial in size because of white-topped aster's rhizomatous habit. Large distances between populations are currently the cause of relative isolation since areas of distribution are not well connected via corridors. Isolation of populations may partially be due to low seedling survival rates, which may be caused by its apparent inability to compete well with other taxa during the seedling stage (Giblin 1997, 96).

This species is fairly intolerant of disturbance (Clampitt 1993, 168) and invasion by exotics (Thomas and Carey 1996, 160). One study predicted white-topped aster more likely to be found where there is a minimum of 32 percent native bunchgrass cover because it establishes in gaps between grass clumps (Clampitt 1993, 167-168). This same study saw the highest rates of coverage in sites that had experienced the least disturbance (Clampitt 1993, 165). Occasionally, white-topped aster will be found under tree canopies such as ponderosa pine and Oregon white oak, but is more abundant in open prairies; encroachment and shading by Douglas-fir no doubt greatly impact this species (Thomas and Carey 1996, 152-153). Another study showed greater occurrence of white-topped aster in habitats comprised of more than 50 percent native taxa (Thomas and Carey 1996, 152). Maintaining an environment with a high ratio of native to exotic taxa with plenty of native bunchgrass that is protected from disturbance is critical to the conservation and restoration strategy for this species.

Hall's Aster:

Hall's aster (*Symphyotrichum hallii*) is a state threatened perennial ranked critically imperiled by NatureServe (Camp and Gamon 2011, 264). This plant is rarely found on the south Puget Sound prairies, but historical herbarium collection records show its range in Washington to have been at least from the island of San Juan to near the Columbia River in low elevation prairies. Data from these records show Hall's aster was found in wet prairie sites as well as dry, mounded locations (Consortium of Pacific Northwest Herbaria 2014). Though very rare in Washington, this species is apparently secure in Oregon (Camp and Gamon 2011, 264).

Hall's aster blooms in late summer and is an important nectar source for butterflies. There are only two relatively recent observations of Hall's aster in the south Puget Sound prairies and little information is available concerning the ecological requirements or sensitivities of the species. The limited information available on the ecology and life history of Hall's aster exemplifies the need for continuing research aimed at determining the best approaches to conservation and restoration in order to prevent the extirpation of rare taxa from Washington.

Animals[1]

Butterflies

The relationship of flowering plants and their insect pollinators is eco-

---

[1]This section on animals was adapted from Washington Department of Fish and Wildlife, "Threatened and Endangered Wildlife in Washington: 2012 Annual Report", specifically from the following sections: Taylor's checkerspot pp. 90-95, Mardon skipper pp. 96-99, streaked horned lark pp. 69-73, and Mazama pocket gopher pp. 106-110. For the complete document visit: www.wdfw.wa.gov/publications/01542/wdfw01542.pdf.

logically important because both plants and pollinators typically depend on each other for survival. Pollinator life cycles are generally closely adapted to the plant communities with which they are associated. For example, adult butterflies are dependent on nectar of certain plant taxa to fuel them through their adult life stage. In the process, nectar-feeding facilitates the transportation of pollen between individual plants. Adult butterflies also choose specific host plant species for oviposition (egg-laying) so that their larvae will have access to the appropriate food source needed for development and survival from hatching to pupating. Although many plant species are capable of self-fertilization and/or asexual reproduction, genetic diversity increases with cross-fertilization, greatly improving the potential for adaptation to changing environmental conditions. For plants with limited ranges and isolated populations, inbreeding can often lead to lowered fertilization, seed development, and germination rates. For many plants, pollination and fertilization is entirely dependent on associated pollinators.

Taylor's Checkerspot:

In 2013, Taylor's checkerspot butterfly (*Euphydryas editha taylori*), was listed as federally endangered after being a candidate species since 2001. The state of Washington has listed this subspecies as endangered since 2006 and NatureServe ranks it as critically imperiled. Once numerous throughout grasslands from southern Vancouver Island along the Salish Sea to the Willamette Valley, Taylor's checkerspot now sustains just seven populations in Washington (Washington Department of Fish and Wildlife 2013).

After emerging from pupation in early spring, adult Taylor's checkerspots live only for a couple of weeks. While seeking a mate during their brief adult life, they survive on the nectar of flowers. They are generalists in their nectar-feeding habits and so may serve as pollinators for a wide range of plant taxa. Observations of nectar-feeding by Taylor's checkerspots have been made for deltoid balsamroot (*Balsamorhiza deltoidea*), sea blush (*Plectritus conjesta*), camas (*Camassia* spp.), several *Lomatium* species, field chickweed (*Cerastium arvense* ssp. *strictum*), wild strawberry (*Fragaria virginiana*), and dandelions (*Taraxacum officinale*) (Cheryl Fimbel, pers. comm.).

After mating, females lay eggs on plants of the figwort family (Scrophulariaceae). A few native species known to be used for oviposition are harsh paintbrush (*Castilleja hispida*), golden paintbrush, marsh speedwell (*Veronica scutellata*), and American brooklime (*V. beccabunga*). Non-native plants are also used for oviposition, including the plantains (*Plantago lanceolata* and *P. major*) and thyme-leaved speedwell (*Veronica serpyllifolia* ssp. *serpyllifolia*). Laboratory research has shown its preference for ovipositing on the two native paintbrush species over the abundant and very common plantains (Aubrey 2013, 26). If appropriate host plants are not available, the newly hatched caterpillars will die (Stinson 2005, 80). When eggs hatch, emerging caterpillars depend on their host plants for food until summer, when they go into an inactive state called diapause, typically burrowing underground to protect themselves from the elements. In late winter they become active again, feeding on a broader range of plants including sea blush, blue-eyed Marys (*Collinsia parviflora* and *C. grandiflora*), and dwarf owl-clover (*Triphysaria pusilla*) before pupating.

Decline of Taylor's checkerspot has coincided with the reduction in population and distribution of those plant taxa that provide for the life-cycle needs of this butterfly. Plant propagation for habitat restoration, combined with butterfly propagation for release of caterpillars and adults, has been successful at three sites (Center for Natural Lands Management 2014c). Many agencies and organizations have collaborated on this effort including, Cascadia Prairie-Oak Partnership, Washington Department of Fish and Wildlife, the Oregon Zoo, Sustainability in Prisons Project (via The Evergreen State College), University of Washington, U.S. Fish and Wildlife Service, Department of Defense and Joint Base Lewis-McChord, Center for Natural Lands Management, Washington Department of Natural Resources, and Washington Natural Heritage Program. The majority of the funding for this effort has come from the Department of Defense.

Mardon Skipper:

The Mardon skipper (*Polites mardon*) is a Washington State endangered species and a federal species of concern. Ranging from Washington to California, this butterfly occupies four sites in the south Puget Sound prairies and others in the southern Cascades of Washington. Mardon skippers are dependent on grassland habitats dominated by native species such as Roemer's fescue, and in the south Puget Sound prairies, they almost always use this species for oviposition and as a larval host (Henry and Schultz 2012). Adults feed on the nectar of many plants including early blue violet (*Viola adunca* ssp. *adunca*), common vetch (*Vicia sativa*), common camas (*Camassia quamash*), prairie lupine (*Lupinus lepidus* var. *lepidus*), spring gold (*Lomatium utriculatum*), western buttercup (*Ranunculus occidentalis* var. *occidentalis*), sea blush, and yarrow (*Achillea millefolium*). Research shows Mardon skippers avoid Scotch broom as a nectar provider (Washington Department of Fish and Wildlife 2013). This weed species, along with tall oatgrass (*Arrhenatherum elatius*) are considered major threats to the Mardon skipper.

Habitat restoration for Mardon skippers is ongoing at Sciurus Wildlife Area. Additionally, survey methods for butterflies are being developed and standardized so that more robust statistical trends can be recorded and analyzed.

Vertebrates

Ecosystems with high species diversity, such as the Puget Sound prairies, include many vertebrate animals that play important roles in the healthy functioning of the system. Habitat loss, especially for breeding, is a primary issue affecting the stability and success of the two vertebrates discussed below. For rare taxa, low genetic diversity caused by decreased numbers of individuals within breeding ranges has detrimental effects on community viability. Major declines in population-level genetic diversity can lead to low fertilization, birth rates, and offspring survival.

Streaked Horned Lark:

The streaked horned lark (*Eremophila alpestris strigata*) is a federally threatened subspecies of horned lark that is listed as endangered in the state of Washington. NatureServe ranks the streaked horned lark as critically imperiled in its breeding phase. Estimates as of 2010 indicate the entire population of this subspecies to be fewer than 2,000 individuals, with about 290 birds breeding in all of Washington (Altman 2011).

This subspecies is migratory and although little is known about its winter locality, it is suspected that the majority of the population over-winters south of Puget Sound, possibly on the southern Oregon coast or in the Willamette Valley (Pearson and Altman 2005, 5-6). Endemic to western Oregon and Washington, it once commonly bred on the south Puget Sound prairies, but general loss of habitat has greatly decreased its population. Streaked horned larks build ground nests in grasslands and areas with little vegetation such as airports and fallow agricultural fields. Currently, streaked horned larks in the south Puget Sound are primarily found at the Olympia and Shelton airports and Joint Base Lewis-McChord.

Although dietary preferences of streaked horned larks have not been well studied, adult horned larks in general feed mainly on grass seeds, with some forb seeds and plant sprouts, during both the winter and breeding season. Chicks are raised solely on insects (Stinson 2005, 58).

Many efforts to expand and restore habitat and increase genetic diversity have been completed and others are currently underway. The limited genetic diversity and low egg-hatching rate of the streaked horned lark led scientists to transfer eggs from nests in Oregon to nests at Joint Base Lewis-McChord.

From this experiment, originating with 12 eggs, five or seven young were fledged and at least one male paired, nested, and produced offspring that fledged (Center for Natural Lands Management 2014b).

Other tactics have been employed to aid in the survival of this subspecies, such as: trials enclosing nests in order to limit predation, tilling of land to provide bare ground habitat, removing invasive species (especially those that alter habitat structure, such as Scotch broom), and developing a guide for agricultural land managers to benefit the streaked horned lark. Burning on the prairies has also been shown to improve habitat by decreasing grass thatch and moss cover, opening necessary bare ground for nesting. Multiple agencies and organizations are working together to secure the survival of this endemic bird, including U.S. Fish and Wildlife Service, Washington Department of Fish and Wildlife, Center for Natural Lands Management, Joint Base Lewis-McChord, Oregon Department of Fish and Wildlife, Oregon State University, The Evergreen State College, Department of Defense, Willapa Wildlife Refuge, and Washington State Parks.

Mazama Pocket Gopher:

In our region, the Mazama pocket gopher (*Thomomys mazama*) is comprised of several subspecies (in the south Puget Sound prairies: *T. m. tacomensis* [likely extinct], *T. m. glacialis*, *T. m. yelmensis*, *T. m. tumuli*, and *T. m. pugetensis*). These four extant subspecies of the Mazama pocket gopher are listed as threatened in Washington and federally under the Endangered Species Act (Center for Natural Lands Management 2014a).

Broadly, the range of the Mazama pocket gopher spans western Washington, western Oregon, and a small area of northern California (Stinson 2005, 22). Like Oregon white oak, it is considered a keystone species because its ecological activity affects the greater ecosystem, having radiating impacts on many other plants and animals. An individual pocket gopher can turn as much as three to seven tons of soil per acre each year (Stinson 2005, 28). Burrowing activity changes soil structure and chemistry, while underground food stores and excrement enrich the soil. On the surface, soil excavations create prime conditions for seed germination of many prairie plants, and the presence of Mazama pocket gophers may even increase plant species diversity on the Puget Sound prairies. One report found that the rare white-topped aster benefitted from the activity of gophers (Stinson 2005, 29). Burrows also provide habitat for many amphibians, reptiles, other mammals, and invertebrates such as insects and spiders. Mazama pocket gophers feed on roots and aerial parts of many plants including camas, and are known to eat fungi, disseminating spores in the process. Most prairie plants undoubtedly have important beneficial relationships with fungal species and the Mazama pocket gopher may influence the abundance and distribution of these fungi (Washington Department of Fish and Wildlife 2014).

Because the habitat needs for basic survival, but not necessarily breeding, of the Mazama pocket gopher are often met even in highly degraded or developed prairies, private land owners sometimes find they have populations living on their lands. State law (RCW 77.15.130) protects state and federally listed animals on private property, so property owners may be required to have areas they would like to clear or develop surveyed (Washington Department of Fish and Wildlife 2014).

Efforts to protect and recover populations of Mazama pocket gophers are ongoing. Translocation of gophers into protected areas with good habitat and breeding potential has expanded some established populations at Wolf Haven International and West Rocky Prairie. However, successful translocation is not easy to achieve and requires large numbers of individuals (more than 100 each year) captured and relocated over several years. These sites must also be managed to protect against encroachment by Scotch broom and other shrubs, as the presence of Mazama pocket gophers is negatively associated with a shrubby ecosystem structure.

Conclusions

Long-term environmental destruction of Puget Sound prairie-oak ecosystems has contributed to an increase in rare and compromised taxa in Washington. Both professional researchers and citizen scientists alike have contributed enormously to efforts aimed at protecting this endangered landscape. Driven by groups as diverse as state and federal agencies, universities, non-profits, and volunteers, these contributions have been critical to maintaining this source of irreplaceable biodiversity. Much more investigation and experimentation is currently underway and will be needed to better learn how to support this rare and fascinating ecosystem.

Because habitat fragmentation poses unique challenges to restoration activities, land-owning citizen scientists have the potential to play a key role in supporting prairie restoration. Concerned private property owners on historical and current prairie soils may be in a unique position to help connect fragmented prairie sites by increasing corridors among isolated patches of remaining prairie. This could be achieved through protection and/or restoration of portions of individual private properties for habitation by native prairie plants and animals.

Restoration Ecology
Sarah Hamman

Over the past 150 years, the structure, functioning, and extent of South Sound prairies have been altered by a wide range of disturbances. Fire exclusion has led to invasion of non-native species, encroachment by shrubs and trees, deep thatch accumulations, and extensive moss cover (Foster and Shaff 2003; Hamman et al. 2011). Intensive grazing and agriculture have impacted vast acreages of prairie (Crawford and Hall 1997), leading to introduced pasture grasses and altered soil structure and nutrient content. Urban and suburban development have increased the cover of impervious surfaces and non-native horticultural species across the prairies. The cumulative effect of these disturbances is a highly fragmented, invaded matrix of low quality prairie lands distributed throughout their original range (Crawford and Hall 1997). Because of this, several prairie obligate species are in decline, some having recently been listed by the U.S. Fish and Wildlife Service as endangered or threatened (USFWS 2012). Due to the rarity of the ecosystem as a whole and the dramatic decline in native plant, invertebrate, bird, and mammal species, restoration goals have been designed to address both ecosystem-level structure and functioning and species-specific habitat needs (Thorpe and Stanley 2011). Managers have determined that the primary goals driving restorative actions in South Sound prairies are to: 1) reduce cover and influence of non-native species, 2) build or restore a diverse, resilient native prairie community, 3) provide the habitat needed by target listed species (Taylor's checkerspot butterfly, streaked horned lark, Mazama pocket gopher) throughout their range, and 4) restore natural processes that sustain this diverse prairie habitat (fire regimes, pollination, etc.). The restoration process used by practitioners to achieve these goals in the South Sound prairies generally involves three stages: invasive removal, site preparation, and native enhancement. Each of these stages involves several strategies that work to overcome site- and species-specific challenges.

Invasive Removal

Non-native, invasive species represent one of the greatest threats to prairies throughout the Pacific Northwest (PNW). Altered fire regimes, past land uses, and increased 'edges' from fragmentation have allowed for and, in some cases, enhanced invasion by several noxious non-native species (Table 2). These invasive species have had a huge ecological and economic impact on PNW ecosystems. It is estimated that Scotch broom alone causes over 40 million dollars in forestry losses and control expenses in Oregon each year

Table 2. Non-native invasive species targets with prairie restoration efforts.

Common Name	Scientific Name	Life Form*	State Classification†	Invasive Strategy	Removal Methods
tall oatgrass	Arrhenatherum elatius	PG	not listed	rhizomes, high seed production	broad-spectrum & grass-specific herbicides
Scotch broom	Cytisus scoparius	PS	B	allelopathy, altered soil nutrients, high seed production, long-lived seeds	repeat mowing; prescribed fire; herbicide; hand-pulling
laurel spurge	Daphne laureola	PS	B	vegetative spread	herbicide; hand-pulling
common teasel	Dipsacus fullonum	BF	C	high seed production	herbicide; hand-pulling
leafy spurge	Euphorbia esula	PF	B	high seed production, long-lived seeds, vegetative spread	prescribed fire; herbicide; biological control
mouse-eared hawkweed	Hieracium pilosella	PF	B	high seed production, stolons, allelopathy	herbicide
hairy cats ear	Hypochaeris radicata	PF	C	early sprouting, high seed production	prescribed fire; herbicide
oxeye daisy	Leucanthemum vulgare	PF	C	high seed production	prescribed fire; herbicide
reed canary grass	Phalaris arundinacea	PG	not listed	rhizomatous spread	herbicide & mowing; herbicide & prescribed fire
sulfur cinquefoil	Potentilla recta	PF	B	high seed production, vegetative spread	herbicide
tansy ragwort	Senecio jacobaea	AF, BF, or PF	B	vegetative spread	prescribed fire; herbicide; biocontrol

*A=Annual, B=Biennial, P=Perennial, G=Grass, F=Forb, S=Shrub. †According to Washington State Noxious Weed Control Board (www.nwcb.wa.gov): A = Goal of complete eradication, landowners are required to prevent seeding and can be held financially responsible for cost of treatment; B = Goal of containment and eventual eradication; C = Control is recommended.

(Hulting et al. 2008). Efforts employed to remove these species include prescribed fire, chemical methods (broad-spectrum, broadleaf-specific, or grass-specific herbicide application), mechanical methods (including mowing and hand-pulling), and biological methods (biocontrol via insects or grazing mammals). These treatments have been applied independently and in combination over time at various prairie sites throughout the South Sound. The most effective treatment combinations depend on the species being treated.

Site Preparation

Because most PNW prairies have a long history of invasive species and other disturbances, it is often necessary to prepare sites for native enhancement before planting or seeding. This may involve reducing moss and leaf litter on the soil surface, altering the soil nutrients and/or restoring the structural and functional components of the soil microbial communities.

Prescribed fire is often used to remove decades of accumulated litter and moss in PNW prairies. This tool can be used over thousands of acres each year and, depending on the timing and ignition patterns of the burn, can provide heterogeneous effects for both fire-sensitive (many invertebrates) and fire-adapted (most prairie plants) species across the landscape (Hamman et al. 2011; Martin and Hamman *in press*). In western Washington, prescribed fire is applied in the late summer to fall to remove senesced litter and expose bare soil for seeding. During this window, most native plants are dormant and rare butterflies are burrowed belowground or nestled in root crowns in diapause. This timing provides the greatest ecological benefit while limiting negative impacts to native plant, invertebrate, and bird species. In addition to removing litter and moss, fire also releases butenolides, compounds in plant-derived smoke that promote seed germination of fire-adapted species (Flematti et al. 2004). Elliott (*in press*) found that butenolides infused in water (smoke water) increased germination rates for some native prairie species and had no effect on others. This technique continues to be studied to determine its applicability to PNW prairie species. Finally, depending on intensity, prescribed fire can either reduce soil nitrogen (through nitrogen volatilization in smoke in higher intensity burns) or increase soil nitrogen (through ash deposition in lower intensity burns) (Giovannini and Lucchesi 1997), which can favor natives or non-natives, respectively.

Due to the invasion of Scotch broom, a nitrogen-fixing leguminous shrub, the biogeochemical cycling of most PNW prairies has been altered (Haubensak and Parker 2004; Caldwell 2006), leading to additional non-native invasions (Suding et al. 2004). One method that is being investigated to restore low-nutrient prairie soils and limit non-native invasion is carbon addition (sugar, sawdust, or charcoal) (Perry et al. 2010). Surface carbon application or soil incorporation (depending on edaphic conditions) can reduce available soil nitrogen and phosphorus through microbial immobilization of these nutrients, which creates conditions that favor slow-growing native species. This method has been tested in grasslands worldwide (Perry et al. 2010), and has shown promising results thus far in PNW prairies (Kirkpatrick and Lubetkin 2011; Mitchell and Bakker 2011).

An additional soil amendment that is currently under investigation in South Sound prairies is mycorrhizal inoculation. Because the prairies have been largely dominated by Scotch broom and non-native grasses that have their own distinct mycorrhizal associates (Börstler et al. 2006; Grove et al. 2012), there may be a microbial legacy that could influence the trajectory of the aboveground communities (Rook et al. 2011). Inoculating native plants with beneficial microbial communities before outplanting could make for an easier transition to the harsh prairie environment. Mycorrhizal fungi can enhance survival and growth of newly reintroduced native species by assisting with nutrient and water uptake and pathogen resistance (Hoeksema et al. 2010). Initial results of a mycorrhizal inoculation study found positive impacts on the initial growth of five and survival of four south Puget Sound species (Porter 2014).

Native Enhancement

Once the invasive species and the limiting factors (thick moss and litter, altered soil nutrients, antagonistic microbial communities) have been removed or amended, native seeds or seedlings can be restored to the site. In areas slated for butterfly reintroduction (currently <50 acres), managers have chosen to transplant native plants into dense patches to provide immediate oviposition and nectar resources. In other prairie restoration areas (currently >500 acres), native seeding is used to provide both food and nesting resources for the streaked horned lark and the Mazama pocket gopher. This method results in a slower native community development, as it can take 2-3 years for some prairie species to flower. A wide suite of other prairie invertebrate, bird, and mammal species benefit from these native habitat enhancements.

Native plant enhancement across hundreds to thousands of acres requires a reliable source of native propagules that have been appropriately sourced to either maintain genetic integrity or enhance genetic diversity. Native plant and seed production in the south Puget Sound has been growing for the past ten years, with 2013 production totaling nearly 500,000 plugs and over 1000 lbs of seed of 65 native species. These plants and seeds are produced through partnerships between the Center for Natural Lands Management, the Sustainability in Prisons Project, Joint Base Lewis-McChord, and several regional nurseries and seed farms. Because most of these species are not produced horticulturally, propagation protocols are not available. As native prairie species production has expanded over the past decade, partners and students have been developing protocols and recording them in the Native Plant Network's Propagation Protocol Database (www.nativeplantnetwork.org/network/) for others to access and utilize for future production.

Due largely to the altered fire regime, the abundance and richness of native annual species on the prairies has dropped considerably over the past century (Dunwiddie et al. 2014). This fact, and the important butterfly resources provided by several native annual species, has encouraged nursery managers to ramp up production of native annuals. Annual species now account for nearly 40 percent of the regional native prairie seed production, up from less than five percent in 2010. This should translate to a much higher abundance and diversity of native annual species present in South Sound prairies in upcoming years.

Achieving diverse, resilient native prairie communities that support rare prairie flora and fauna requires a complex set of restoration strategies to address both the above- and belowground structural and functional components of this system. The collaborative and multi-faceted restoration effort in the South Sound prairies aims to do just this.

History of White Settlement

Lisa Hintz

In 1853 and 1854, the United States federal government funded the scouting of the best route for a railroad between the Mississippi River and the Pacific Ocean (Messmer 1994, vi). As a part of this effort, the War Department subsidized an investigation into the natural history of the land. Two of the men involved were Isaac Stevens, then governor of Washington Territory, and James Graham Cooper, a surgeon and naturalist (Messmer 1994, vi). Cooper's voluminous works based on the railroad survey include two botanical documents: "Report Upon the Botany of the Route" and "Catalogue of Plants Collected in Washington Territory." In his reports, he identified 150 species "peculiar" to the prairies, not present in the surrounding forest (Cooper 1859, 15).

Cooper's descriptions are important to our current and historical understanding of the prairies as an anthropogenic landscape, and of widespread Euro-American perspectives on the land that rendered many, including Cooper, incapable of fully understanding their origins. In his descriptions of the dry prairies of the Upper Chehalis River in the spring of 1854, Cooper wrote:

> Here the gravelly soil characterizing the whole valley between the Coast and Cascade ranges, together with a drier climate, had produced…a very distinct group of flowers still blooming in abundance, made it seem as if we had in the distance of a few miles reached an entirely new country…I recognized at once the characteristic plants of the dry prairie near Vancouver and the Cathlapoot'l River [a tributary of the Columbia River, now named the Lewis River], where the preceding summer I noted, in July, that "we passed through, in a distance of fifty miles, seven prairies from one to four miles in width…forming a charming contrast to the almost impenetrable forests." (13-14)

And like many early Euro-American naturalists, he recognized the prairies as cultural landscapes:

> A few remarks are necessary upon the origin of the *dry* prairies so singularly scattered throughout the forest region. Their most striking feature is the abruptness of the forests which surround them, giving them the appearance of lands which have been cleared and cultivated for hundreds of years. The Indians, in order to preserve their open grounds for game, and for the production of their important root, the camas, soon found the advantage of

burning... (16; emphasis in original)

However, as well as he seemed to understand the existence of the prairies as a product of the management activities of their Indigenous human inhabitants, Cooper seemed to have trouble grasping the potential for people to have a creative, rather than purely destructive, impact on their environments:

> With all this magnificence there is not wanting scenery of a milder and more home-like aspect. The smooth prairies, dotted with groves of oak, which in the distance look like orchards, seem so much like old farms that it is hard to resist the illusion that we are in a land cultivated for hundreds of years and adorned by the highest art... Nothing seems wanting but the presence of civilized man, though it must be acknowledged that he oftener mars than improves the lovely face of nature. (34)

Had Cooper been able to understand that the prairies were, in fact, maintained as gardens, for hundreds, even thousands of years by Native peoples, he may have been better able to consider the capacity of humans to interact with nature in positive ways.

Since the early days of European and Euro-American contact with the Salish people of the Puget Sound region, the anthropogenically maintained prairies were, in terms of their ecological functioning and wild food production capabilities, alternately appreciated, sought after, and destroyed. A relatively rapid series of events led to the ultimate transformation of the prairies in the past 200 years or more, beginning with the loss of Indigenous lives and traditional societal functioning, and continuing with white settlement and its associated agricultural, industrial, and military enterprises.

Disease

Smallpox likely came to the Puget Sound around 1782 via the routes of British, French, and Russian fur traders. George Vancouver witnessed the devastating toll of European diseases during his exploration of the area in 1793 (White 1980, 26-27). These diseases were a biological nightmare for the Native peoples, who did not understand their origins and had little immunity. By 1840, roughly half the Native population of Puget Sound had died (White 1980, 27). Such a massive loss of lives severely harmed the structure and functioning of communities, and subsequently impacted the anthropogenic prairie-oak landscapes. The reduction and eventual cessation of burning and other land-use practices, both in frequency and spatial scale, resulted in rapid encroachment by Douglas-fir and, thus, the subsequent shrinkage of one of the major food resource areas of the Indigenous human population (Leopold and Boyd 1999; White 1980, 26).

Puget Sound Agricultural Company

Between the late eighteenth and mid-nineteenth centuries, explorers and scouts ventured through the Puget Sound, but direct settlement did not come about until the British-owned Hudson's Bay Company established a fort near the mouth of the Nisqually River in 1833 (Troxel 1950, 17). By the 1840s, a significant white farming population lived in the south Puget Sound area (White 1980, 36). Many of these farmers worked for the Hudson's Bay Company, which began operating the Puget Sound Agricultural Company in 1839 with support from Fort Vancouver to the south. This subsidiary was formed to produce meat (mainly beef and mutton) for the multitudes of trappers and traders, both non-Native and Native, who were employed in the region by the Hudson's Bay Company (Troxel 1950, 136-137). The company also produced agricultural commodities to meet contractual agreements with the Russian American Company, another fur business operated out of Russia via Siberia, through the Aleutians into present-day Alaska and southward as far as California (Blumenthal 2009, 21; Ficken 2002, 5).

The land claim of the Puget Sound Agricultural Company was approximately 167,000 acres (as well as an additional 3,500 acres on the Cowlitz River) (Gray 1930, 99) and stretched between the Puyallup and Nisqually Rivers, from the Cascade foothills westward to the Puget Sound. Historical land claim maps indicating vegetation from 1855, drawn up by William Fraser Tolmie, who worked for the Hudson's Bay Company, show the area was mostly open prairie or "plain" (Washington Office of the Secretary of State 2014), akin to the landscape near the Chehalis River observed by Cooper during the two previous years. The Hudson's Bay Company records indicate that prairie fires, certainly intentionally set by the Salish, occurred almost annually during this time (Perdue 1997, 19). Such fires likely disturbed the company people at the Nisqually Fort because of the potential harm that could be wrought on their infrastructure. It is possible that the company pressured the Nisqually people to discontinue this land management practice (Perdue 1997, 20) as white settlers did in other regions of the West.

Farmers preferred this open land for agricultural purposes because forested land was difficult and expensive to clear. Charles Wilkes, an American who led the nautical exploration of Washington State's portion of the Salish Sea, described the area near Fort Nisqually in 1840: "[N]othing could be more beautiful, or to appearance more luxuriant than the plains, which were covered with flowers of every colour and kind: among these were to be seen Ranunculus, Scilla [now *Camassia*], Lupines, Collinsia, and Balsamoriza [sic]." He goes on to describe the settlement of the Puget Sound Agricultural Company:

[T]hey have a large dairy, several hundred head of cattle, and among them seventy milch cows…they also have large crops of wheat, peas, and oats, and were preparing the ground for potatoes. These operations are conducted by a farmer and dairyman, brought from England expressly to superintend these affairs. (Blumenthal 2009, 20)

According to Helen Norton (1979), based on an account from Tolmie in 1841,

The Nisqually Plain produced good grazing grass and the Puget Sound Agriculture Company imported thousands of sheep, Spanish 'horned cattle', and horses to feed on this grass. These animals (over 16,000 all together) competed with native animals for forage and the sheep soon ruined all the prairies as sources of food plants. (179)

In the century to follow, imported agricultural land use practices rapidly affected an increasing amount of acreage with the immigration of more people from Asia, Hawaii, Europe, the eastern U.S., and elsewhere.

Early Settlement

The majority of the settlers living and working at the Puget Sound Agricultural Company land claim were British, but in the decades following the opening of the Oregon Trail in 1843, Americans from the east and an ever-increasing number of European immigrants began to tip the balance between British and American residents in the Pacific Northwest, first in the Willamette Valley in Oregon, and eventually in the Puget Sound (Ficken 2002, 6). The primary influence on this population influx was The Preemption Act of 1841, which allowed settlers to buy 160 acres of surveyed land at $1.25 an acre, and by mid-1854, the law included unsurveyed lands as well (White 1980, 37-38).

It cannot be stressed enough how the land-use and economic ideologies imported by European and Euro-American settlers impacted the prairies of the south Puget Sound. Simply put, white settlers believed farming and capitalist accumulation by industrial resource extraction were the most valuable uses of the land. These ideologies are exemplified by communications from the era, such as this description by Clinton A. Snowden in his 1909 book, *The History of Washington: the Rise and Progress of an American State*, which concerns the presence and use of exploitable resources in the Puget Sound during the mid-nineteenth century:

Its rich valleys were uncultivated, its mines unopened, its vast wealth of timber undisturbed… Nature everywhere lavished her prodigal hand, but there was no one to profit by it… [The] fertile valleys and undulating hills...waited only for water and the plow to change them into teeming farms and bending orchards... (421)

An early settler on Whidbey Island named Walter Crockett provided another example from 1853: "[T]o get the land subdued and the wilde *[sic]* nature out of it. When that is accomplished we can increase our crops to a very large amount and the high prices of everything that is raised heare *[sic]* will make the cultivation of the soil a very profitable business" (White 1980, 35). Such ideology in practice ultimately resulted in devastating effects on prairies.

These historical observers may not have seen the land the way Cooper saw it in the same decade, as appearing to have "been cleared and cultivated for hundreds of years" and "adorned by the highest art." Although Cooper could not understand his own observations of the prairie landscape, he saw something of the work and effort of the Native peoples in transforming the land that sustained them. Their transformation was extensive, as would be the Euro-American transformation that had already begun.

Initial American settlement of the south Puget Sound area came in waves, encouraged by cheap land sales and early advertising. The first settlement north of the Columbia River was established by way of the Cowlitz Trail, founded in 1845 at present-day Tumwater, then called New Market. The Hudson's Bay Company had substantially supported its establishment by providing food, money, and a donation of machinery for a sawmill on the falls, all of which allowed for the survival of the settlement for a short time (Ficken 2002, 7). The following year, when the 49th parallel was established as the boundary between United States and British territory, the Hudson's Bay Company was allowed to keep most of the Puget Sound Agricultural Company land claim. Consequently, it maintained a firm hold on the economy and foreign trade in the region (Ficken 2002, 8-9).

In 1850, the Donation Land Law of Washington was enacted. This law granted 320 acres to white or "half-breed Indian" male settlers over the age of 18 who were or intended to become United States citizens, and who had come to Washington Territory before December 1, 1850. Wives (also white or "half-breed Indian") of male settlers received 320 acres in their own names if they had married their husbands before December 1, 1851. Anyone fitting these criteria that arrived between December 1, 1851 and December 1, 1854 was entitled to 160 acres of land. All blacks, Hawaiians, Indians, and Asians were excluded from the land claims. Settlers only had to occupy and cultivate the land for four consecutive years to make the land title final (Bevan 2004).

In truth, until Isaac Stevens, the first governor of Washington Territory, established treaties with the tribes west of the Cascades in late 1854, all of

Washington Territory belonged to the tribes according to federal law, as none had yet been ceded to the United States government. All lands acquired by settlers prior to 1855 via the Donation Land Law and the Preemption Act, as well as the lands held by the British Hudson's Bay Company, were illegal claims (Ficken 2002, 43-46). Additionally, some of the lands allotted under these two acts were not yet surveyed, gridded, and mapped by the General Land Office into the township/range system that had been devised by the Land Ordinance of 1785. Claims by the Hudson's Bay Company caused disputes with American settlers (Carpenter 1986, 56-162; Scott and DeLorme 1988, map 30), and essentially all claims caused dispute with the existing tribes. As this rapid and increasing occupation caused tensions among groups, violent conflicts sometimes broke out (Carpenter 1986, 143-154 and 172-190; Scott and DeLorme 1988, map 28; Ficken 2002, 44-45).

Most of the land granted to settlers in western Washington by the Donation Land Law was likely prairie, as most of the land grants in the south Puget Sound area were located in areas adjacent to current extant prairie (Scott and DeLorme 1988, map 30). Centuries of management by the Salish had kept these lands relatively clear; with a theoretically minimal amount of work, they could be ready for European cultivation practices or livestock grazing. Records show the total number of land grants from the Donation Land Law by county: Lewis, 96; Pierce, 108; and Thurston, 234 (Riddle 2010). By 1854, the region was primarily populated by Americans and less so by British citizens. Most of the settlers worked as subsistence farmers, producing wheat, oats, peas, beans, potatoes, or dairy (Perdue 1997, 23), and keeping urban service businesses.

Boosterism, the convincing advertisement and enthusiastic persuasion aimed at drawing more people to a place, successfully encouraged more settlers to move to the region. In 1849 there were 304 white settlers north of the Columbia River, in the region that would, in 1853, become Washington Territory (Snowden 1909, 443). In 1850, between 1,049 and 1,201 white settlers lived in the same area. Though widely scattered, principal population centers were at the two forts of the Hudson's Bay Company at Vancouver on the Columbia River and Nisqually, and at the head of Budd Inlet (Snowden 1909, 443; U.S. Census Bureau 2014).* By 1860, the settler population grew to just over 3,000 in Thurston, Pierce, and Lewis Counties (U.S. Census Bureau 2014). These censuses included whites only; census taking excluded Native people although they were more numerous than the settlers at this time. In 1853, Isaac Stevens and his colleagues calculated that roughly 7,000 Native people lived in Washington Territory west of the Cascade Range (Ficken 2002, 44).

The 1862 Homestead Act became yet another route by which white

---

* These numbers only include the far southwestern portion of present-day Washington State.

settlers could obtain inexpensive land. This Act declared that any current or intended U.S. citizen who had never fought in battle against the United States government, and who was over the age of 21 or the head of a household (including independent women), was allowed to file a claim on a quarter section of 160 acres (Potter and Schamel 1997). After five years of living on and "improving" the land, including building a small home and growing food crops, a certificate of ownership was granted (Potter and Schamel 1997; Scott and DeLorme 1988, map 31). Although this act was technically open to all races (including freed slaves), laws barring certain nationalities from immigration and citizenship in effect controlled who could participate. However, by this time Donation Land Act claimants had already taken most of the open prairie lands, and the Homestead Act did not greatly further affect the settlement and transformation of the prairie-oak landscapes (Scott and DeLorme 1988, map 31).

The influx of white settlers, most of whom intended, at least initially, to take up farming as a means of survival, brought a variety of changes to the land. However, not all of these changes were intentional, and as settlement increased, so did the ecological impacts on prairie-oak ecosystems.

Impacts on the Prairies by Euro-American Farming Practices

Perhaps the most powerful force of direct, early, and permanent transformation of the prairie was the introduction of European and Euro-American agricultural practices that accompanied initial settlement. As Richard White put it in his 1980 book *Land Use, Environment, and Social Change: The Shaping of Island County, Washington*, "[T]he farmer composed the vanguard of ecological invasion of North America. He introduced to the continent, both intentionally and accidentally, the exotic plants and animals that have permanently altered the natural systems of the New World" (35). This is certainly true for the Puget Sound prairies.

Farmers had immediate impacts on the prairie-oak ecosystems. Of all the species with great value to the Salish, perhaps only the trees were of importance to the settlers (White 1980, 42). Lands were tilled for crops and grazed by livestock, increasing summer evapotranspiration (White 1980, 43) and damaging soil structure, no doubt having radiating effects on microhabitats and pollinators. Both fenced and unfenced domesticated animals, including sheep and cattle, ate the native bunchgrasses down to the ground. Edward Huggins, an agent for the Puget Sound Agriculture Company, commented on the decline of the prairies:

> The nutritious blue bunch grass was plowed up or killed out by too close pasturing and followed the cattle into the things of the past. The most

diligent cultivation failed to make the gravelly soil of the plains produce profitable crops; fields again were turned into pastures which produced scant growth much inferior to the original blue bunch grass which, Huggins says, he has seen waving in the breeze like the great fields of ripening grain. (Norton 1979, 179)

Pigs particularly harmed the prairies because their rooting behavior destroyed plant communities and especially impacted the abundance of camas, which they loved to eat. White's (1980) in-depth descriptions about Island County, although not in the south Puget Sound, can be used as a model representative of the changes in the land wrought to the prairies in the mid-nineteenth century. Presumably in the south Puget Prairie region, Euro-Americans also affected large animal populations, similar to White's descriptions of Island County (49-51). Wolves and cougars were killed to protect livestock, while elk were likely hunted to low numbers, impacting the ecological relationships of the region.

Farmers brought in vast amounts of seed to provide fodder for their animals, which then became tough competition for native species. Introduced plants included oats (*Avena sativa*), wheat (*Triticum*), clover (*Trifolium*), Timothy (*Phleum pratense*), and colonial bentgrass (*Agrostis capillaris*). Many invaders problematic for the farmers themselves tagged along. Such species included Canadian thistle (*Cirsium arvense*), sow thistle (*Sonchus oleraceus*), common velvet grass (*Holcus lanatus*), shepherd's purse (*Capsella bursa-pastoris*) and sheep sorrel (*Rumex acetosella*) (White 1980, 49). During the railroad survey in the early 1850s, Cooper also compiled a list of adventive species (Cooper 1859). With records from such an early date, we can see how rapidly naturalization of non-native species occurred. His list includes currently persistent inhabitants of the south Puget Sound prairies: shepherd's purse, California poppy (*Eschscholtzia californica*), annual bluegrass (*Poa annua*), sheep sorrel, prickly sow-thistle (*Sonchus asper*), and red sandspurry (*Spergularia rubra*). Scotch broom, one of the most detrimental of invasive plant species that became established on the prairies, was introduced to Vancouver Island in the 1850s, and by 1900 it was naturalized on the island (Kaplan 2003). It probably spread to the mainland and the Puget Lowland via the ornamental trade and along highway routes, following Euro-American settlement. Other species, likely brought by settlers intentionally, remain severely problematic to native prairie communities. Examples include St. John's-wort (*Hypericum perforatum*), and oxeye daisy (*Leucanthemum vulgare*) (Buschmann 1997, 166 and 168).

Settlers' trails and roads, and eventually the railroad, were often the same routes traveled by the exotic and invasive species (White 1980, 40 and 46). This disturbance is still the avenue by which many invasive species find their way to new habitats. These plants were not necessarily better adapted to the local environment, but were likely more able to take advantage of the disturbance caused by the migrations of white settlers and their accompanying agricultural practices. Studies have shown that competitive dominance occurs when a species is better able to utilize limited available resources and outcompete other species for these resources. However, contingent dominance might better explain the success of some of these invasive exotic species. Contingent dominance can take at least two forms. For one, the events of human disturbance in concert—here, intensive grazing practices, fire suppression, and the introduction of a suite of pasture grasses and other weeds—"could easily create a dominant and stable exotic assemblage that need not rely on relative competitive superiority" (MacDougall 2002, 160). Or, secondly, exotic species that have been following Europeans around the globe may be better adapted to the disturbance cycles of European land-use practices (MacDougall 2002, 160).

In complete opposition to Native peoples' ecological management strategies for food production, fire was, for the most part, an outright enemy of settlers. Once permanent homes were built and farms had been established, farmers generally feared and prevented fire by whatever means were available. Soon timber also became an economically important resource, ultimately deemed worthy of protection from fires that could spread from the prairies. Fire suppression quickly ushered in fast-growing Douglas-fir to the prairies and associated oak woodlands. Early records by settlers indicate this conifer encroachment to have been noticeable quite soon after fire cessation began. J. Harlan Bretz, an early American geologist, wrote in his 1913 bulletin for the Washington Geological Survey about the encroachment of trees into the prairies: "It is stated that the Indians formerly burned over these prairies annually, and destroyed the trees growing on them. It is certain that today, the forest is encroaching[...] Many gnarled skeletons of the broad-spreading prairie oaks are found mouldering in a dense growth of young fir which has killed them in the last half century" (Bretz 1913, 101-102). Indeed, this encroachment is still advancing today (Kruckeberg 1991, 301).

Although early settlement by white farming communities in the south Puget Sound prairies was swift and affected rapid changes to the land, it was also illegal, according to United States federal law. Salish inhabitants of the area fought for the right to use their lands as they always had, but opposition and conflict from the settlers was strong. By late 1854, the legal standing of the United States to possess the majority of the land in the Puget Sound was finalized.

Official Tribal Loss of Land by Treaty

On December 26, 1854, dozens of western Washington tribes and Governor Isaac Stevens signed the Treaty of Medicine Creek (Ficken 2002, 46). By this treaty, the tribes ceded most of their lands for little more than $30,000, small reservations, and the theoretical right to hunt, fish, and gather on traditional grounds. In 1855, following the signing of the Treaty of Medicine Creek, Native peoples were sent to reservations with devastating results to both their societies and their traditional prairie homelands (White 1980, 19). In concert, the cessation of frequent, low-intensity fires, the relatively easy settlement of open areas, the predominating European land-use ideology that privileged domesticated plants and animals over wild species, and the biological invasion of diseases and weeds resulted in the swift, vast reduction of the anthropogenic prairie landscape. In effect, prairie-oak communities were destroyed over most of their former range. During this era of change, early industrialization was moving in from the east and south.

The Railroad

The Northern Pacific Railroad, among other smaller railroad enterprises, was in 1862 granted by Congress an alternating checkerboard of one-square-mile parcels of land from the Great Lakes to the Pacific Ocean (Schwantes 1989, 144). This land grant occupied the entirety of the south Puget Sound, although lands that had already been taken by settlers were not granted to the railroad companies. Finances, politics, and the sheer immensity of such an endeavor impeded the progress of building the railroad. Still, by 1890 several railroad lines ran through the lowlands of the South Sound. The Tacoma, Olympia, and Gray's Harbor Railroad connected on both ends to tracks owned by the Northern Pacific, linking Montesano in the west and Tacoma in the east. The Northern Pacific built a line further south, connecting Tenino to Tacoma and Seattle (Scott and DeLorme 1988, map 47).

The construction of railroads damaged the south Puget Sound prairie-oak cultural landscapes and neighboring ecosystems. Enormous amounts of lumber were needed for railroad ties and vast acreages of forest were cleared to produce them and replace them when they rotted. Lands where the track was to be laid as well as adjacent areas had to be cleared and leveled. Few people considered the ecological impacts of railroad construction and maintenance, or the impacts of the growing population arriving via the new transportation infrastructure.

Mining, especially for gravel, was also a growing industry during this time, and was instrumental in the building of the railroad. Because of the geologic history of the south Puget Sound lowlands and the effects of the last glaciation on the soils and topography, this region contains extensive gravel deposits. Deep deposits accumulated primarily in the lowlands during the post-glacial floods (Moen 1986, 2-3 and map 3) and provide one of the key hydrologic foundations for the formation of the prairies. There are dozens of former and extant gravel mines located within the historical reach of the last glaciation. Gravel mining operations no doubt have impacted prairie ecology and contributed to the fragmentation of existing habitat.

Urbanization

With the coming of the railroad, the late nineteenth century saw a rise in the number of settlers to urban centers of the Puget Sound (White 1980, 64). With this population influx came an increase in size and complexity of developed urban areas. Extractive industry such as logging and mining increased, partly as expanding urbanization itself necessitated greater resource extraction, which also provided employment for growing numbers of laborers. Eventually nearly all of the historical prairie lands in the Pacific Northwest were overtaken. Continued development still negatively affects the prairies today, especially in many rural areas with remnant prairie communities.

Fort Lewis

Established in 1916 by an 86 percent vote in Pierce County, Fort Lewis today occupies over 86,000 acres to the east of the Puget Sound between Olympia and Tacoma in both Pierce and Thurston Counties. Development of the fort displaced hundreds of white settlers along with their livestock and agricultural practices, and this left some original, yet somewhat transformed, prairie relatively intact (Perdue 1997, 23-24). While 16,210 acres of the original 66,225 acres had been invaded by or planted with Douglas-fir since 1870, small pockets of prairie and oak savanna remain (Tveten 1997, 123). Today, Fort Lewis is home to two-thirds of south Puget Sound's extant prairie (Goodrich 2013). Contemporary usage of Fort Lewis is not limited to military activities alone. Various recreational uses and timber harvesting, as well as traditional practices of the Nisqually and Chehalis Tribes occur on the fort (Schmidt 1997, 262).

Damage to the prairies has occurred on the base, as it has been a military training ground involving tanks, trenches, foxholes, large arms fire, parachute drops, and artillery targets (Goodrich 2013). Tracked and Stryker vehicles in particular pose a major threat to the plant and animal communities of the prairie and oak savanna at Fort Lewis. Although a frequent fire regime is important to the ecology of the prairies, fires intentionally or accidentally set each year (likely too frequent) by military training in the Artillery

Impact Area may cause harm by reducing the diversity and populations of native species. Although the Salish may have burned prairies annually, variable patchiness of large burns may have prevented areas from experiencing the full effects of fire disturbance on an annual basis (Hamman et al. 2011, 321). Current restoration fire frequency goals in the Puget prairies are set at three to four years (Hamman et al. 2011, 321). It should be noted that some of the prairies at Fort Lewis are considered the most intact relative to their presumed historical species composition, and this may be due, in part, to the frequent fires that have continually been set since the fort was established as a training ground.

While some activities at the fort have great potential to damage what is left of these relatively highly intact prairie-oak ecosystems, Fort Lewis simultaneously maintains a protective relationship with this endangered habitat. The first protections were established in 1994 from a Final Environmental Impact Statement aimed at minimizing damage to prairie ecology (Perdue 1997, 25). Early prairie restoration at Fort Lewis used prescribed burning on a three to five year basis covering a total of 7,413 acres (Tveten 1997, 124). Today, the U.S. Department of Defense's (DoD) Legacy Program is one of the largest contributors to prairie restoration and conservation in Washington State. The program has supported seed and propagule production of native species as well as research into protocols for germination and growth (Goodrich 2013). The burden of prairie conservation and restoration falls heavily on the U.S. Army because Joint Base Lewis-McChord hosts some of the few remaining populations of three federally listed species – Taylor's checkerspot butterfly, streaked horned lark, and Mazama pocket gopher – including the only site in the world where all three species exist together. The DoD has a vested interest in both maintaining open areas for military training and maintaining habitat for these species under the regulations of the Endangered Species Act (Goodrich 2013).

Conclusions

Prairie communities likely comprised just ten percent of the south Puget Sound landscape at the time of European arrival (Crawford and Hall, 13). Assessments of soil type and aerial photos, combined with estimates produced using various sources such as historical maps, public land records, and ethnographic accounts tell us that only about three percent of the anthropogenically-maintained prairie ecosystem still exists today in the south Puget Sound. One assessment estimated the transformation of prairie as follows: 33 percent to urban development, 32 percent to forest or other ecological conversion, and 30 percent to agriculture (Crawford and Hall 1997).

Due to changes in prairie-oak community composition, driven by shifts in global climate occurring roughly 10,000 years ago, Native peoples' implementation of extensive burning, and the consequences of settlement and European and Euro-American land-use ideologies, the prairies of the south Puget Sound have been in long-term flux. In the 180 years since white settlement, changes have occurred rapidly, resulting in significant loss of these cultural landscapes. Invasion of exotic plants, an overall decrease in community and genetic diversity due to habitat loss and fragmentation, changes in soil composition by Douglas-fir encroachment and nitrogen fixation from Scotch broom, and continued urban, suburban, and agricultural development all impact the remaining prairies.

Though these negative impacts can seem impossible to overcome, many people and agencies are working to restore and maintain the remnant, extant prairies in the region. Personal reasons for doing this work are varied; however, Patrick Dunn, a long-time prairie restoration ecologist and director of the South Sound Prairies Program with the Center for Natural Lands Management, provides a compelling argument for prairie conservation beyond protecting plant and animal diversity in the Puget Lowland. He brings together the immediate physical reality of the land with our historical knowledge and perceptions, both Native and non-Native:

> The remnants of prairie that exist today are reminders of our heritage, a physical link to our ancestors and how they perceived the Puget Sound. The act of prairie restoration respects both biological and human qualities. By reestablishing or creating habitats, restoration supports the plants, animals, and ecological processes of the prairies. Restoration also links people to their heritage by actively involving them in preserving that heritage, the prairie landscape of the south Puget Sound. (Dunn 1998, 4)

That heritage can be found in the traditional life-ways of the Salish peoples of this land, from their frequent fires, the camas harvests, the use of other edible and medicinal plants from the prairie, and more. That heritage is also clearly seen in the writings of the white settlers and explorers, like James Cooper, who saw these lands as unique, profoundly altered by human hands, but could not fully appreciate the anthropogenic nature of so "magnificent" a landscape. While he could see with his eyes the beauty and "home-like aspect" of the prairies through which he traveled, he could not see beyond the lens of his cultural ideology that these lands were, in fact, a type of farm, in full cultivation, for an entire society. Revealingly, Cooper candidly conceded the inevitable ill effects of "man" (albeit, as he understood man) on the "lovely face of nature."

Although white settlement of the Puget Sound drastically changed the landscape within a very short time, and the same land-use ideologies that informed those changes are still at work today, rising environmental awareness has inspired strong efforts to prevent further degradation and to restore the ecological, cultural, and aesthetic nature of the prairies.

Holistic Ecological Restoration

Rose Edwards

Many restorationists embrace the Society for Ecological Restoration's (SER) definition of restoration, which is: "the process of assisting the recovery of an ecosystem that has been degraded, damaged, or destroyed" (Apostol 2006, 11). Fostering native plant and animal biodiversity is often critical in this work. Because our prairies, savannas, and oak woodlands are cultural landscapes maintained prior to white settlement by Native people, the involvement of current day Native communities is important. In the south Puget Sound region, restoration strategies often include the Native practice of frequent low intensity burning. However, holistic restoration that engages Native communities in restoring prairies, savannas, and oak woodlands has been limited, including restoration and re-creation of rituals and traditions that tie one's physical and spiritual wellbeing to the health of a landscape.

In *Nature by Design*, Eric Higgs examines American cultural myths that romanticize nature and settler takeover of lands (46). He writes about how we fetishize the idea of "authenticity" while also wanting nature to stay under our ultimate control, carefully manicured (54). The common duality between humans and nature pervades our thinking, and Higgs invites us to consider how it and other common cultural beliefs influence various approaches to restoration work (94).

Restoration should not be a form of ecological nostalgia, because all landscapes are cultural landscapes and cultures continue to live and change (Higgs 2003, 93-94). Unfortunately, most Americans lack sufficient ecological literacy that integrates honest accounts of the human history of where they live (Higgs 2003, 56). The avoidance of our colonial legacy has real consequences. We have collectively recreated the history of the landscapes we call home in ways that erase stories of widespread human engagement with the land. The loss of Native American land stewardship mirrors the loss of healthy prairie-oak ecosystems. Direct Euro-American management of the land further damaged prairie-oak woodland ecosystems through fire suppression, overgrazing, row cropping, development, and the introduction of invasive species (Hosten et al. 2006, 65).

Native people have lived sustainably on this land for thousands of years. Tending and harvesting camas and other food plants created and still creates a bond between people and the prairies, which nourishes both in deep, sustaining ways. Restoration will work best when whole communities become involved in sustainable land stewardship. Indigenous leadership is ideally a component of restoration, but there are many written accounts that can be used to educate the non-Native public even in the absence of partnership. How can we re-mediate if we don't truly understand the history of what happened? Simply through example, renewed Indigenous cultural practices can aid in the restoration and management of the land (Higgs 2003, 235).

Indigenous-led camas field restoration is already occurring just to the north of the Puget Sound. Tribal historian and land manager Cheryl Brice of the Songhees First Nation works to rejuvenate camas and its use on Discovery Island and Chatham Island off the coast of Vancouver Island, British Columbia. Through her work with the community, Cheryl is renewing the land's ability to foster camas production, then harvesting and cooking the resulting bulbs using traditional methods such as pit roasting (Senos et al. 2006, 416-417). Her people numbered in the thousands in this region for centuries, but by 1911 fewer than one hundred survivors remained (Higgs 2003, 229). Despite the losses, Cheryl's connection to the land remains strong as she works to prevent the loss of "yet another model for how to live in a place" (Higgs 2003, 230). According to Higgs, "the past offers wise counsel but no simple lessons" (239). The same can be said of the land, which holds the impressions of history.

Many Native peoples want to revive tradition, to strengthen their communities and create greater cultural continuity. The Songhees First Nation's effort to reclaim camas production on Discovery Island and Chatham Island shows us how the process of reviving camas harvest using traditional knowledge and traditional practices can also incorporate modern tools and methods. A traditional digging stick remains the best way to harvest camas, but a number of conveniences, such as lighters and inflatable boats, may be used to arrive at the celebratory meal that comes after harvesting. This is a good example of where restoration is being led by Native people. This type of restoration, which combines ecological integrity, a strong sense of history, and a commitment to the land, has been called 'focal restoration' (Senos et al. 2006, 416-417; Higgs 2003, 206-244). Dennis Martinez, the Chair of SER's Indigenous Peoples' Restoration Network, distills his vision of the restoration that is needed when he says: "What do we want to restore? We want to restore life. We want to restore the living and sacred relationship between the people and the earth. We want to restore our spirits as we restore the land. We want to restore our culture, our songs, our myths, and stories, and the Native names for creeks and springs. We want to restore ourselves" (Higgs 2003, 121).

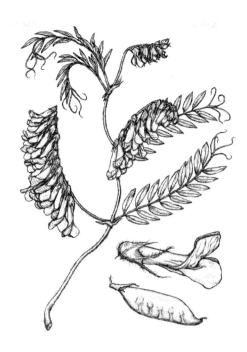

Plant Illustrations & Descriptions

The illustrated taxa in this field guide represent 149 of the vascular plants that occur in the prairies and associated oak woodlands of the south Puget Sound region. Nomenclature and family treatments followed are primarily that of Hitchcock and Cronquist (1973). Where they are not, synonyms are provided in brackets. For current nomenclature, we have relied on the online Washington Flora Checklist. In the second edition of this field guide, the nomenclature and family treatments will be updated after the revised flora for the Pacific Northwest is published. In addition to increasing the number of taxa illustrated, the vouchered list of vascular plants in Appendix A will be expanded significantly. Fifteen students and Evergreen faculty member Frederica Bowcutt generated the illustrations from field sketches primarily at Glacial Heritage County Park and Scatter Creek Wildlife Area. They also drew from herbarium specimens housed at Evergreen's herbarium within the college's Natural History Museum. For a complete list of the illustrators, refer to Appendix B. For an analysis of the vascular plant flora of the South Sound Prairies of Washington, refer to Dunwiddie et al. (2006).

List of Illustrated Plants

PTEROPHYTA | FERNS
Polypodiaceae Polypody Family
Polypodium glycyrrhiza licorice fern
Pteridium aquilinum var. *pubescens* bracken fern

CONIFEROPHYTA | CONIFERS
Pinaceae Pine Family
Pinus contorta var. *contorta* shore pine
Pinus ponderosa var. *ponderosa* ponderosa pine
Pseudotsuga menziesii var. *menziesii* Douglas-fir

ANTHOPHYTA | FLOWERING PLANTS
Dicotyledoneae—Plants with two seed leaves
Apiaceae [Umbelliferae] Carrot Family
Daucus carota wild carrot, Queen Anne's lace
Lomatium nudicaule barestem biscuitroot, pestle parsnip
Lomatium triternatum var. *triternatum* nineleaf biscuitroot
Lomatium utriculatum spring gold
Sanicula crassicaulis var. *crassicaulis* Pacific sanicle, blacksnake root

Apocynaceae Dogbane Family
Apocynum androsaemifolium spreading dogbane

Asteraceae [Compositae] Aster Family
Achillea millefolium yarrow
Agoseris grandiflora var. *leptophylla* Puget Sound agoseris
Anaphalis margaritacea western pearly everlasting
Antennaria howellii ssp. *howellii* [*A. neglecta* var. *howellii*] field pussytoes
Balsamorhiza deltoidea deltoid balsamroot
Crepis capillaris smooth hawksbeard
Erigeron speciosus showy fleabane
Eriophyllum lanatum var. *leucophyllum* Oregon sunshine, common woolly sunflower
Gaillardia aristata blanket flower, common gaillardia
Hieracium albiflorum white-flowered hawkweed
Hieracium scouleri Scouler's hawkweed

Hypochaeris radicata   hairy cat's ear
Leucanthemum vulgare   oxeye daisy
Microseris laciniata ssp. *laciniata*   cutleaf microseris
Sericocarpus rigidus [Aster curtus]   white-topped aster
Solidago missouriensis   Missouri goldenrod
Solidago simplex var. *simplex [S. spathulata* var. *neomexicana]*   Mount Albert goldenrod, coast goldenrod
Tragopogon dubius   yellow salsify
Taraxacum officinale   common dandelion
Wyethia angustifolia   mule's ears

Berberidaceae   Barberry Family
Berberis aquifolium   tall Oregon-grape

Boraginaceae   Borage Family
Myosotis discolor   yellow and blue forget-me-not
Plagiobothrys scouleri var. *scouleri*   Scouler's popcorn flower

Brassicaceae [Cruciferae]   Mustard Family
Capsella bursa-pastoris   shepherd's purse
Cardamine oligosperma   little western bittercress, shotweed
Draba verna   spring draba, spring whitlow-grass
Lepidium campestre   field pepperweed, pepperwort, field cress
Teesdalia nudicaulis   barestem teesdalia, shepherd's cress

Campanulaceae   Harebell Family
Campanula rotundifolia   common harebell, bluebell-of-Scotland
Triodanis perfoliata   clasping Venus'-looking-glass

Caprifoliaceae   Honeysuckle Family
Sambucus racemosa var. *racemosa*   red elderberry
Symphoricarpos albus var. *albus*   common snowberry
Viburnum ellipticum   common viburnum, oval-leaved viburnum

Caryophyllaceae   Pink Family
Cerastium arvense ssp. *strictum*   field chickweed, mouse-ear

Cucurbitaceae   Cucumber Family
Marah oregana   wild cucumber, coastal manroot

Ericaceae   Heath Family
Arctostaphylos uva-ursi   kinnikinnick

Fabaceae [Leguminosae]   Pea Family
Acmispon parviflorus [Lotus micranthus]   small-flowered deer vetch
Cytisus scoparius   Scotch broom, Scot's broom

Lupinus albicaulis   sicklekeel lupine, white-stemmed lupine
Lupinus bicolor   two-color lupine, miniature lupine, small-flower lupine
Lupinus lepidus var. *lepidus*   prairie lupine
Lupinus polyphyllus var. *pallidipes*   bigleaf lupine
Trifolium dubium   least hop clover, suckling clover
Trifolium pratense   red clover
Trifolium subterraneum   burrowing clover
Vicia americana var. *americana*   American vetch
Vicia sativa var. *angustifolia*   common vetch
Vicia villosa var. *villosa*   hairy vetch

Fagaceae   Beech Family
Quercus garryana var. *garryana*   Oregon white oak

Geraniaceae   Geranium Family
Geranium molle   dovefoot geranium

Hypericaceae (now Clusiaceae)   St. John's-wort Family
Hypericum perforatum   common St. John's-wort

Lamiaceae [Labiatae]   Mint Family
Clinopodium douglasii [Satureja douglasii]   yerba buena
Lamium purpureum   dead-nettle
Prunella vulgaris   self-heal, heal-all

Onagraceae   Evening Primrose Family
Chamerion angustifolium [Epilobium angustifolium]   fireweed
Clarkia amoena var. *lindleyi*   farewell-to-spring

Orobanchaceae   Broomrape Family
Orobanche uniflora   naked broomrape

Plantaginaceae   Plantain Family
Plantago lanceolata   narrowleaf plantain, English plantain
Plantago major   common plantain, great plantain, nippleseed
Plantago patagonica   hairy plantain, Indian wheat

Plumbaginaceae   Plumbago Family
Armeria maritima ssp. *californica*   thrift, sea-pink

Polygonaceae   Buckwheat Family
Rumex acetosella   sheep sorrel

Portulacaceae   Purslane Family
Claytonia perfoliata [Montia perfoliata in part*]*   miner's lettuce

Primulaceae Primrose Family
Dodecatheon hendersonii Henderson's shooting star, broad-leaved shooting star
Dodecatheon pulchellum var. *pulchellum* dark-throated shooting star

Ranunculaceae Buttercup Family
Anemone lyallii Lyall's anemone
Aquilegia formosa var. *formosa* western columbine
Delphinium nuttallianum upland larkspur
Ranunculus occidentalis var. *occidentalis* western buttercup
Ranunculus uncinatus little buttercup

Rhamnaceae Buckthorn Family
Frangula purshiana [Rhamnus purshiana] cascara sagrada

Rosaceae Rose Family
Amelanchier alnifolia serviceberry
Crataegus douglasii black hawthorn
Crataegus monogyna one-seed hawthorn
Fragaria virginiana wild strawberry
Oemleria cerasiformis Indian plum, osoberry
Potentilla gracilis var. *gracilis* slender cinquefoil
Rosa gymnocarpa bald-hip rose
Rubus spectabilis salmonberry
Rubus ursinus trailing blackberry, dewberry

Rubiaceae Madder Family
Galium aparine cleavers, bedstraw, goosegrass
Galium parisiense wall bedstraw
Sherardia arvensis blue field madder

Salicaceae Willow Family
Salix scouleriana Scouler's willow

Saxifragaceae Saxifrage Family
Lithophragma parviflorum small-flower woodland star, small-flower prairie star
Micranthes integrifolia [Saxifraga integrifolia] swamp saxifrage, whole-leaf saxifrage
Tellima grandiflora fringe cup

Scrophulariaceae Figwort Family
Castilleja hispida var. *hispida* harsh paintbrush
Castilleja levisecta golden paintbrush

Collinsia grandiflora   large-flowered blue-eyed Mary, giant blue-eyed Mary, blue-lips blue-eyed Mary
Collinsia parviflora   small-flowered blue-eyed Mary, maiden blue-eyed Mary
Nuttallanthus texanus [Linaria canadensis var. *texana]*   blue toadflax
Parentucellia viscosa   yellow glandweed
Triphysaria pusilla [Orthocarpus pusillus]   dwarf owl-clover
Veronica arvensis   corn speedwell

Urticaceae   Nettle Family
Urtica dioica ssp. *gracilis*   stinging nettle

Valerianaceae   Valerian Family
Plectritis congesta   seablush, rosy plectritis

Violaceae   Violet Family
Viola adunca ssp. *adunca*   early blue violet
Viola glabella   Stream violet
Viola praemorsa var. *praemorsa [V. nuttallii* var. *p.]*   upland yellow violet

Monocotyledoneae—Plants with one seed leaf

Cyperaceae   Sedge Family
Carex inops ssp. *inops*   long-stolon sedge

Iridaceae   Iris Family
Iris tenax   Oregon iris, tough-leaf iris
Sisyrinchium idahoense var. *idahoense*   blue-eyed grass

Juncaceae   Rush Family
Juncus bufonius var. *bufonius*   toad rush
Luzula comosa var. *laxa*   Pacific woodrush

Liliaceae   Lily Family
Brodiaea coronaria   crown brodiaea, harvest brodiaea
Camassia leichtlinii ssp. *suksdorfii*   large camas
Camassia quamash   common camas
Dichelostemma congestum [Brodiaea congesta]   ookow, forktooth ookow, northern saitas, congested snake lily
Erythronium oregonum ssp. *oregonum*   white fawn lily
Fritillaria affinis [F. lanceolata]   chocolate lily
Lilium columbianum   tiger lily
Trillium parviflorum   small-flowered trillium
Triteleia grandiflora [Brodiaea howellii]   Howell's brodiaea
Triteleia hyacinthina [Brodiaea hyacinthina]   white brodiaea
Zigadenus venenosus   meadow death-camas

VASCULAR PLANTS OF THE SOUTH SOUND PRAIRIES 65

Orchidaceae   Orchid Family
Spiranthes romanzoffiana [S. r. var. *r]*   hooded ladies'-tresses

Poaceae [Gramineae]   Grass Family
Agrostis capillaris [A. tenuis]   colonial bentgrass
Aira caryophyllea var. *caryophyllea [A. c. in* part*]*   silver hairgrass
Aira praecox   little hairgrass
Anthoxanthum odoratum   sweet vernalgrass
Arrhenatherum elatius   tall oatgrass
Bromus carinatus var. *carinatus*   California brome
Bromus sterilis   barren brome, poverty brome
Danthonia californica   California oatgrass
Danthonia intermedia   timber oatgrass
Elymus glaucus ssp. *glaucus*   blue wildrye
Festuca roemeri [F. idahoensis var. *roemeri]*   Roemer's fescue
Holcus lanatus   common velvet grass
Koeleria macrantha [K. cristata]   prairie Junegrass
Lolium multiflorum   Italian ryegrass
Lolium perenne   perennial ryegrass
Melica subulata var. *subulata*   Alaska oniongrass
Panicum capillare ssp. *capillare [P. c.* in part*]*   witchgrass, common panicgrass
Panicum occidentale (now *Dichanthelium acuminatum* ssp. *fasciculatum*) hairy panicgrass
Phleum pratense   Timothy
Poa annua   annual bluegrass
Poa pratensis ssp. *pratensis*   Kentucky bluegrass

PTEROPHYTA | Ferns
Polypodiaceae—Polypody Family

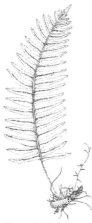

Polypodium glycyrrhiza
licorice fern

The licorice fern is a native perennial that can grow up to 70 cm long, but it is often shorter. The central stipe of the symmetrical blade is yellowish, with pinnae that are bright green. They are most often seen growing on the trunks of big-leaf maples (*Acer macrophyllum*), but they also grow on moist ground, rocks, or logs. The scaly rhizome tastes sweet, like licorice or angelica.

Pteridium aquilinum* var. *pubescens
bracken fern

A deciduous, rhizomatous fern that is densely covered with hairs on frond undersides, bracken fern is usually around 1 m tall, but can grow to 2 m. Fronds are 2 to 3 times pinnatifid with pinnae opposite. Lower pinnae are triangular, upper are lanceolate. Marginal sori are obscured by rolled leaf margins. Coast Salish Tribes have traditionally used the carefully prepared rhizomes for food, the leaves for pit-oven construction, and the rhizome fibers for tinder. *Pteridium aquilinum* is the most widespread fern in the world.

CONIFEROPHYTA | Conifers
Pinaceae—Pine Family

Pinus contorta* var. *contorta
shore pine

Shore pine is an evergreen tree reaching up to 30 m tall. Needles are dark green, typically paired in groups of 2 per fascicle and range in length from 3 to 6 cm. Mature trees feature thin, scaly bark, gray to dark brown in color. This pine is one of our most broadly distributed conifers, with this variety growing near sea-level and another at timberline. The shape varies greatly by habitat; forest trees will generally feature straight trunks, while coastal trees often take on a twisted, wind-shorn appearance. Female (seed) cones are woody, 4 to 6 cm in length, with a sharp prickle at the tip of the cone scales. Female cones often remain on the tree for years after opening. The reddish male (pollen) cones are typically 1 cm in length and are borne in clusters near the ends of branches.

Pinus ponderosa var. *ponderosa*
ponderosa pine

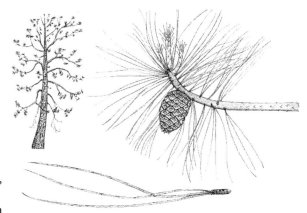

As Washington's only native pine with needles in bundles of 3, ponderosa pine is fairly easy to identify. The needles are 12 to 25 cm long. An oddity in western Washington, ponderosa pine grows in the droughty soils on Joint Base Lewis-McChord. Washington's largest ponderosa is about 65 m tall and grows south of Mt. Adams in the Gifford Pinchot National Forest. Adapted to the presence of fire, ponderosa pine features thick, scaly bark and can survive low-intensity fires that do not reach the crown.

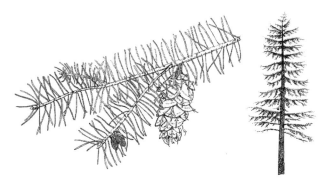

Pseudotsuga menziesii var. *menziesii*
Douglas-fir

Douglas-fir is a large conifer, reaching up to 90 m tall, with needle-shaped, evergreen leaves. Mature trees have thick, deeply furrowed, dark brown bark. Seed bearing cones feature distinctive 3-lobed bracts between the cone scales. Considered to be invasive on South Sound prairies, Douglas-fir can rapidly colonize areas where fire has been excluded. Douglas-fir encroachment also threatens savanna and woodland species such as Oregon white oak (*Q. garryana* var. *garryana*) and western gray squirrel (*Sciurus griseus*). In western Washington, shade intolerant Douglas-fir trees only naturally regenerate following stand clearing disturbances such as fires or blowdowns.

ANTHOPHYTA | Flowering Plants

Apiaceae [Umbelliferae]—Carrot Family

Daucus carota
wild carrot, Queen Anne's lace

This weedy herb introduced from Europe and naturalized throughout the U.S. is the "parent" plant of the garden carrot. Wild carrot is biennial and can grow to 120 cm tall. Its leaves are highly dissected, giving it a lacy appearance. The tiny flowers are borne in compound umbels that are cream to pink in color, often with a purple or pink central flower. It can be found at low elevations, in disturbed areas such as roadsides, fields, and clearings. It blooms in July and August.

Lomatium nudicaule
barestem biscuitroot, pestle parsnip

Blooming from April to June in dry, open areas, this native perennial herb grows from an enlarged, edible taproot, reaching a height of 20 to 90 cm. Leaves are glaucous and 1 to 3 times compound. Leaflets are lanceolate to ovate. Small, yellow flowers are borne in open compound umbels supported by unequal rays to 20 cm. Often the peduncles are swollen and hollow at the base of the umbel.

Lomatium triternatum var. *triternatum*
nineleaf biscuitroot

Nineleaf biscuitroot is a native perennial herbaceous plant that grows up to 80 cm tall, blooming from April to July. It has small yellow flowers in somewhat loosely aggregated compound umbels. The leaves are mostly basal or lower cauline, ternately to ternate-pinnately compound, with narrow leaflets and sheathed bases. It can be found in low to mid-elevations with dry to moist soils. In addition to inhabiting the South Sound prairies, this food plant can also be found in the sagebrush steppe of eastern Washington.

Lomatium utriculatum
spring gold

This common lomatium is a native perennial herbaceous plant that grows to 60 cm tall, blooming April through June. It has lacy, highly dissected leaves with obvious sheathing at the base. Flowers are yellow and arranged in compact, compound umbels. Bracts are slightly toothed and located at the base of each umbellet rather than at the base of the compound umbel. It is found at low elevations, in open, rocky meadows and slopes, often growing with grasses.

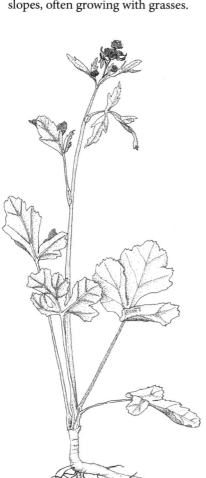

Sanicula crassicaulis var. *crassicaulis*
Pacific sanicle, blacksnake root

This is a native perennial carrot relative with an aromatic taproot. It grows 25 to 120 cm tall and has palmate leaves with 3 to 5 sharply defined lobes. Lower leaves are petiolate; upper leaves are often sessile. Small yellow flowers, sometimes with a purple cast, are borne in irregular compound umbels. Umbels occur in clusters or singly and are subtended by leafy bracts. Fruits are oval burs with hooked prickles. Pacific sanicle grows at low elevations in moist to dry soils, in open forest or habitat edges where it blooms from April to June.

Apocynaceae—Dogbane Family

Apocynum androsaemifolium
spreading dogbane

Spreading dogbane is a many-branched native perennial herb with opposite, more or less oblong leaves. Reaching 50 cm in height, it bears bell shaped flowers with 5 pinkish lobes in terminal and axial cymes. This plant is fairly common on drier sites from low to high elevations. Spreading dogbane is used by Native Americans as a fiber plant for basket making. It blooms from June to September.

Asteraceae [Compositae]—Aster Family

Achillea millefolium
yarrow

Yarrow is a common perennial herb growing up to 1 m in height from a rhizome. Both native and introduced populations are found in the Pacific Northwest. The leaves are pinnately dissected, appearing fringed, reaching 5 to 20 cm long. The leaves are largest at the base of the plant, and smallest near the top. The terminal inflorescence is somewhat flattened, with flowers ranging from white to pink in color. Yarrow can be found in a variety of habitats across the northern hemisphere, where it has been used medicinally for centuries by numerous peoples. It blooms from April to October.

Agoseris grandiflora var. *leptophylla*
Puget Sound agoseris

The large-flowered agoseris is a native perennial that prefers dry open areas such as the prairies. Growing 20 to 70 cm tall, it has yellow dandelion-like flowers that emerge from May to August. They are borne singly and begin with the head bent towards the ground, gradually straightening as they mature. The involucre is composed of strongly triangular bracts, giving the unopened bud a distinctive appearance. Leaves are oblanceolate to linear, 10 to 25 cm long, entire or irregularly pinnatifid. The leaves are much narrower than those of dandelions and form basal rosettes, with no cauline leaves.

Anaphalis margaritacea
western pearly everlasting

This native perennial herb is widely distributed across Washington and occurs throughout most of North America. It inhabits dry to somewhat moist open areas at low to mid-elevations. Western pearly everlasting is dioecious, meaning that male and female blossoms are borne on separate individuals. It grows upright from rhizomes, 20 to 90 cm tall. The stems are white and woolly. The lanceolate to narrowly oblong or linear leaves are woolly as well, though less so above. The leaves are sessile, deep green on the upper surface, often curled under, and up to 12 cm long by 2 cm wide. The composite flower heads, up to 1 cm in width, are numerous and crowded onto a broad, somewhat flattened inflorescence. Pearly-white bracts, globose in bud, enclose the flower heads and are dry and papery, making the inflorescences easy to preserve by drying - hence the common English name. It blooms from July through September.

Antennaria howellii ssp. *howellii*
[*A. neglecta* var. *howellii*]
field pussytoes

A native perennial herb, field pussytoes bears small, scaly heads composed of inconspicuous white disk florets. The basal leaves are egg-shaped to lanceolate. Leaves borne on vertical stems (15 to 40 cm tall) are sparse and narrow. This dioecious plant spreads vegetatively by stolons. Sexual reproduction by field pussytoes may be limited in prairies as male plants are relatively rare in our area. It blooms from April to June.

Balsamorhiza deltoidea
deltoid balsamroot

This native perennial herb has a large woody taproot and showy yellow inflorescences resembling sunflowers. The bright green triangular-shaped basal leaves grow on long petioles and reach 20 to 60 cm in length. Multiple terminal inflorescences with bright yellow ray florets and darker disk florets are borne on stalks 20 to 90 cm long. Deltoid balsamroot can be found in open areas blooming from March through July.

Crepis capillaris
smooth hawksbeard

Smooth hawksbeard is usually an annual but can be biennial. This European introduction grows mainly in disturbed soil in open areas in our region. Milky juice, or latex, exudes when stems or leaves are cut. Inflorescence comprised of 20 to 60 ray florets, each with a fused corolla which is toothed on the outer edge and 8 to 12 mm long. There are several inflorescences on one stem. Leaves are lanceolate to oblanceolate with fine teeth of irregular length. Leaves decrease in length from the base to the top. The whole plant is 10 to 90 cm high. It blooms from May through October.

Erigeron speciosus
showy fleabane

This showy native perennial herb grows up to 75 cm tall. The lower leaves are strap-like and pointed to spatulate, the upper leaves often lanceolate and sessile. Tiny yellow disk florets form the dense center of each inflorescence and are surrounded by light purple ray florets which appear to form the "petals" of a false flower or pseudanthium. Flower heads cluster at the terminus of the plant. Showy fleabane can be found in open clearings and foothills up to middle elevations, blooming from June through August.

Eriophyllum lanatum var. *leucophyllum* [E. l.]
Oregon sunshine, common woolly sunflower

Oregon sunshine is a silvery-woolly native perennial herb, usually with multiple stems per plant. Leaves are usually opposite, sometimes alternate, deeply lobed to entire. Yellow inflorescences of both disk and ray florets are borne singly on stems reaching 10 to 60 cm in height. Ray florets are bright yellow, usually with deeper color towards the center. It prefers dry and open areas and blooms from May to August.

Gaillardia aristata
blanket flower, common gaillardia

Generally found east of the Cascades, blanket flower is a native perennial herb, often with a woody base persisting throughout winter. Stems can be solitary to several, with plants reaching 20 to 70 cm in height. Leaves are narrow, cauline and (occasionally wholly) basal, with entire to toothed or nearly pinnatifid margins. Leaves, including the petiole, grow to 15 cm by 2.5 cm wide and, along with the stems, are hairy. The inflorescence is both ligulate and discoid, the disk from 1.5 to 3 cm wide and dark purple or brown. Ray florets are generally 13 in number, 3-lobed at the tips, 1 to 3.5 cm long, and yellow, becoming purplish towards the base. Heads emerge from long peduncles, either singly or a few to an individual. This species blooms across seasons from May to September.

Hieracium albiflorum
white-flowered hawkweed

This native perennial member of the aster family features white ray florets, blooming from June to August. The florets are subtended by an involucre of greenish-black bracts. Basal leaves are sparsely hairy, while upper leaves are smooth. Leaf outline varies with some leaves featuring wavy margins, while other leaf margins are entire. White-flowered hawkweed can be occasionally found on the margins of prairies.

Hieracium scouleri
Scouler's hawkweed

Both the leaves and the stem of Scouler's hawkweed are covered in fine hairs, with denser hairs near the base of the plant. Leaves are elliptical, growing progressively smaller as you near the inflorescence. An involucre of overlapping gland-tipped bracts subtend the yellow ray florets. This native perennial hawkweed grows in open habitats on both sides of the Cascades, blooming from June to August.

Hypochaeris radicata
hairy cat's ear

This weedy introduced perennial herb is commonly seen in prairies, lawns, disturbed areas, and along roadsides. The yellow inflorescence is composed of ray florets only and are borne atop stems growing 15 to 60 cm tall. The stems are often branched and contain latex. Its entire to lobed leaves are all basal, very hairy and prostrate to the ground. It blooms from May to October and produces copious quantities of achenes with downy pappuses, much like dandelions.

Leucanthemum vulgare
oxeye daisy

This common, weedy non-native perennial herb grows in disturbed areas, meadows, and fields. Oxeye daisy inflorescences consist of white ray florets, 1 to 2 cm long, surrounding a 1 to 2 cm cluster of yellow disk florets. Terminal heads are borne singly, but individual plants may branch, having multiple heads. Basal leaves are spoon-shaped, with cauline leaves becoming more lanceolate and serrate to deeply lobed further up the stalk. Stems are ribbed and often have reddish longitudinal stripes. It blooms from May to October.

Microseris laciniata ssp. *laciniata* [M. l.]
cutleaf microseris

This interesting and relatively common native plant is found in open areas from the eastern foothills of the Cascades to the coasts of Washington, Oregon, and California. It is a perennial herb with mostly basal, entire to deeply laciniate-pinnatifid leaves which are often mottled. Flowers emerge singly on long stems, buds face downwards and turn up towards the sun upon blooming. The involucre is 12 to 25 mm in length, and flowers are yellow with solely ligulate florets. Achenes are 5 to 6 mm and lack beaks. Cutleaf microseris blooms from May through July.

Sericocarpus rigidus [*Aster curtus*]
white-topped aster

A native perennial herb that spreads by rhizomes, white-topped aster produces unbranched stems 10 to 30 cm in height. The lanceolate leaves are 2.5 to 3.5 cm long, trending smaller up the stem towards the inflorescence. White-topped aster is considered threatened in Oregon and listed as a sensitive species in Washington, due, in part, to prairie invasives such as Douglas-fir and Scotch broom. It blooms in July and August.

Solidago missouriensis
Missouri goldenrod

This native perennial herb with hairless stems grows from 20 to 90 cm from creeping rhizomes. Leaves are alternate and lanceolate in form. Terminal clusters of small yellow flowers form long narrow panicles. Each pseudanthium, or inflorescence, is made up of 7 to 13 ray florets and 8 to 13 disk florets. The fruits are small achenes with a downy pappus. Missouri goldenrod is distributed throughout the US and prefers dry, open places like roadsides and prairies from low to high elevations. It blooms from late June to October.

Solidago simplex var. *simplex*
[*S. spathulata* var. *neomexicana*]
Mount Albert goldenrod, coast goldenrod

A native perennial occupying most of the western United States and much of Canada, Mount Albert goldenrod is highly variable and can be found in open areas with moist soil from alpine meadows to coastal dunes. This species is glabrous and grows from 10 to 40 cm with mostly basal leaves arising from a short, woody stem. Cauline leaves become reduced and sessile further up the stalk. Basal leaves are up to 15 cm long including the petiole and up to 3 cm at the widest point. They are oblanceolate to spathulate, hence the Latin binomial, with entire to toothed margins. Inflorescences are borne on panicles that can be elongate to short and compact. Involucres are 4 to 6 mm, ray florets are usually 4 to 10 in number, disk florets number 10 to 18 and both are yellow. This species blooms from June to September.

Taraxacum officinale
common dandelion

Dandelion is a taprooted perennial herb introduced from Europe that grows up to 30 cm high. The leaf ranges in length from 6 to 40 cm, is deeply toothed, and mostly glabrous. The lobes are sharp and angular, unlike the rounded lobes of hairy cat's ear. Terminal inflorescences are borne singly on hollow scapes containing latex. They are composed of ligulate flowers that are bright yellow. These ray florets are subtended by an involucre of bracts in two series. Fruits are achenes with parachute-like pappuses borne in a spherical head. The achenes are dispersed by wind. Dandelion is a cosmopolitan weed that thrives in disturbed areas and lawns. It blooms from March to October.

Tragopogon dubius
yellow salsify

Introduced from Europe, yellow salsify has pale yellow ray florets, with only one ligulate inforescence per stem. Leaves are lance-like, attached to the stem at swollen nodes, sometimes with fine hairs in their axils. Involucral bracts exceed ray florets in length. This annual to biennial plant ranges 30 to 100 cm in height and blooms from May to August. By late summer yellow salsify produces a large and showy, spherical head of bristled achenes that travel on the wind. It is found in drier open areas, chiefly east of the Cascades.

Wyethia angustifolia
narrow-leaved mule's ears

This fragrant, native perennial herb grows from a taproot in meadows and open areas at low elevation. Leaves are usually simple. Basal leaves have large, elongate blades tapering at each end with short petioles, while cauline leaves are smaller, alternate, and often sessile. Leaves and stems are pubescent and involucral bracts are ciliate on the margins. Flower heads are usually solitary with light yellow disk florets. Darker yellow ray florets number from 13 to 21 with ligulate blades 1.5 to 3.5 cm long. Mule's ears bloom from May to July.

Berberidaceae—Barberry Family

Berberis aquifolium
tall Oregon-grape

Tall Oregon-grape is a native evergreen shrub with glossy, prickle-tipped pinnate leaves, which are divided into 5 to 9 leaflets. Reaching 2 m in height, tall Oregon-grape grows in open and edge habitats. Yellow flowers with 6 petals are borne in a branched inflorescence 3 to 8 cm tall. The dark purple-blue fruits of Oregon-grape are edible, with a strongly sour flavor. Tall Oregon-grape can be distinguished from dull Oregon-grape (*B. nervosa*) by counting the leaflets. Only dull Oregon-grape will have more than 9.

Boraginaceae—Borage Family

Myosotis discolor
yellow and blue forget-me-not

One of many naturalized species of *Myosotis*, this annual or short-lived perennial grows 10 to 40 cm tall. The plant is hairy, with fibrous roots and alternate, lanceolate leaves. Terminal spike inflorescences curl into "fiddleheads." The small flowers start yellow and age to blue, seldom blooming below the midpoint of the stalk. This distinguishes it from *M. laxa*, which has showier flowers that are only blue. The corolla is fused at the base to form a narrow tube, then spreads abruptly into 5 lobes. It is a low elevation plant that grows near roads and in meadows, blooming from late April through August.

Plagiobothrys scouleri var. *scouleri*
Scouler's popcorn flower

A small, annual native herb, Scouler's popcorn flower can be found in open, moist areas of the prairie. The flowers grow in tight, elongate clusters and have 5 petals fused at the base with a small set of appendages in the throat of the corolla where the petals become yellow. The sepals are also fused at the base and very hairy. Leaves are also hairy, narrow and opposite towards the bottom of the plant, sometimes becoming alternate up the stem. Species of this genus are sometimes identified by their seeds, which have a ventral keel and a scar. Scouler's popcorn flower blooms in early spring.

Brassicaceae [Cruciferae]—Mustard Family

Capsella bursa-pastoris
shepherd's purse

Shepherd's purse is an annual herb introduced from Europe that has become widely distributed in disturbed ground throughout North America. The simple or branched stems, with tiny stiff hairs, are 10 to 50 cm tall and grow from a rosette of shallowly to deeply lobed basal leaves. The cauline leaves are alternate, sessile and clasping, with a lanceolate to oblong-oblanceolate shape and remotely serrate edges. From March to July, many-flowered open racemes of tiny 4-petaled flowers bloom from the bottom upward, resulting in a progression of fruit along the vertical axis. The fruits are strongly flattened triangular-obcordate silicles, like tiny green hearts with somewhat pointed lobes.

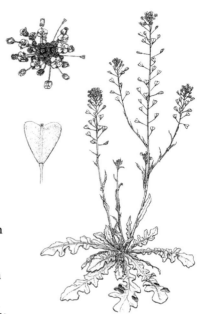

Cardamine oligosperma
little western bittercress, shotweed

This common and widely distributed little native mustard can be annual or biennial and makes a spicy addition to a salad when picked from an uncontaminated site. It grows in a variety of habitats, but prefers wet areas. Little western bittercress produces one to several stems from a central taproot, reaching 10 to 40 cm in height. The stems are usually freely branched and sometimes have short, stiff hairs. The pinnately compound leaves form a basal rosette; cauline leaves are alternate and resemble the basal leaves. They are composed of 4 to 10 lateral leaflets, obovate to ovate or sub-orbicular in shape with a few rounded lobes and one considerably larger terminal leaflet. White flowers are borne from March to July in several-flowered racemes, usually bractless, but sometimes with bracts on lower flowers. The fruit is an erect linear silique, 1.5 to 2.5 cm long, containing 15 to 32 seeds. They dehisce ballistically, giving the plant one of its common names.

Draba verna
spring draba, spring whitlow-grass

One of the earliest plants to bloom, spring draba makes its debut in February and continues to bloom through May. It can be very small, growing from 2 to 20 cm tall with leafless stems that may or may not be branched. The leaves are all basal, simple, narrow, with weak teeth on the margin and branched hairs on the surface. The flowers are small and white, with 4 deeply notched petals, borne in open terminal racemes. The fruit is an elliptic silique. Widely distributed throughout Washington and the rest of North America, this annual herb was introduced from Europe and now enjoys circumboreal distribution. It tends to be found in shrub-steppe, grasslands, and opened disturbed areas at low to middle elevations.

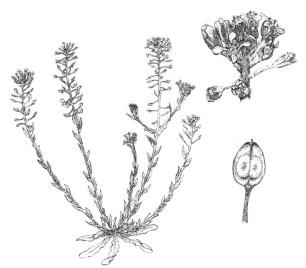

Lepidium campestre
field pepperweed, pepperwort, field cress

This Eurasian annual to biennial herb is distributed across Washington and through much of North America in scattered locations. Field pepperweed has a somewhat grayish appearance and grows in dry, disturbed areas. The simple or distally branched stems range in height from 10 to 60 cm. The basal leaves form a rosette and have dense short hairs. The shape of the basal leaves is variable, from entire to pinnately lobed. The numerous alternately arranged cauline leaves are oblong-lanceolate in shape with toothed edges and sessile, clasping bases. The small, white, 4-petaled flowers appear in May to June in simple or compound slightly flattened racemes with spreading pedicels. Fruits are oblong-ovate silicles with margins and tips that are winged and upturned.

Teesdalia nudicaulis
barestem teesdalia, shepherd's cress

Another member of the mustard family introduced from Europe, barestem teesdalia is an annual herb that grows in sandy or gravelly soil at low elevations in western Washington and Oregon. Its oval to oblanceolate or obovate leaves can be entire to deeply lobed. Most leaves are part of the basal rosette. The plant has a few cauline leaves. Simple or freely branched stems rise 5 to 25 cm from the base, ending in a terminal bractless raceme of tiny 4-petaled white flowers in April and May. The fruits are strongly obcompressed silicles, oblong-obovate in shape, 3 to 3.5 mm long and nearly as wide. They appear somewhat heart-shaped, though less triangular and more egg-shaped than Shepherd's purse (*Capsella bursa-pastoris*), with which it is sometimes confused.

Campanulaceae—Harebell family

Campanula rotundifolia
common harebell, bluebell-of-Scotland

Common harebell is a native perennial rhizomatous herb, growing 10 to 80 cm tall. As one of its common names suggests, this is a circumboreal species. Basal leaves are round to heart-shaped and die before flowering. Cauline leaves are linear and slender. The nodding, bell-shaped flowers are bluish-purple and borne in a lax raceme. It can be found in open, rocky areas from low elevations to alpine areas, blooming from June to September.

Triodanis perfoliata
clasping Venus'-looking-glass

A close relative of common harebell, clasping Venus'-looking-glass is an upright, branched or unbranched annual growing to 30 cm. The alternate, clasping, round to somewhat heart-shaped leaves are rough to the touch. Flowers are sessile and emerge singly or in small clusters directly above the leaves. Lower flowers are self-fertile and do not bloom, while middle and upper flowers are light to deep purple with 5 partially fused petals with lobes longer than the corolla tube beneath. The 5 sepals are also fused at the base and very pointed at the tips. A relatively common native species, Venus'-looking-glass blooms from mid-spring to mid-summer.

Caprifoliaceae—Honeysuckle Family

Sambucus racemosa var. *racemosa [S. r.]*, red elderberry

This deciduous native shrub grows up to 3 to 6 m tall and has soft brown bark. The pinnately compound leaves have 5 to 7 lanceolate to oval leaflets with finely serrated edges. Panicles of white to cream flowers are borne from March to July and develop into red berries relished by birds. Red elderberry grows rarely in the prairies, preferring shade. Blue elderberry is more common in the prairies.

Symphoricarpos albus var. *albus*
common snowberry

Common snowberry is a small erect, native deciduous shrub with opposite branching that usually reaches about 2 m in height. Growing from rhizomes, it tends to form large thickets. Leaf margins can be smooth or wavy, occasionally lobed on new growth. Small campanulate flowers, 5 to 7 mm long, are borne in terminal clusters from May to August. Snowberry produces clusters of white fleshy fruits, 1 to 1.5 cm long, that provide food for birds and mammals throughout the winter. Snowberry can grow in dry or moist soils, in open forests, ravines, and slopes, at low to mid-elevations.

Viburnum ellipticum
common viburnum, oval-leaved viburnum

A deciduous shrub growing to 3 meters in height, oval-leaved viburnum can be found in lowland habitats from southern Washington to northern California. Branching is opposite, as are the broadly elliptical leaves, each with 2 thin stipules at the petiole-stem axils. The inflorescence is an umbel of white 5-lobed flowers, blooming May through June. Edible, fleshy fruits are red and single-seeded, maturing to become shiny black in the fall.

Caryophyllaceae—Pink Family

Cerastium arvense ssp. *strictum* [*C. a.*]
field chickweed, mouse-ear

This showy native perennial herb forms tufts or loose mats. The plant ranges from hairless to glandular-pubescent, with narrow opposite leaves 1 to 3 cm long and 1-nerved. Open inflorescences of 3 to 5 white flowers with 5 bi-lobed petals, 1 to 3 cm long, are borne atop stems from 5 to 50 cm high. The fruits are membranous, slightly curved cylindrical capsules. Field chickweed grows in open areas from low-elevation to subalpine. It blooms from April to August.

Cucurbitaceae—Cucumber Family

Marah oregana
wild cucumber, coastal manroot

This native perennial cucumber relative trails or climbs to considerable heights with winding tendrils. It has striking, alternate, palmately lobed leaves that are rough and hairy to the touch and cordate at the base. The campanulate white flowers are either male or female, but reside on the same plant. The bladder-like fruits are sparsely to densely covered with flexible prickles that harden with age. When fruits dry, they burst open at the apex. It can be found in moist open soils at low elevations, blooming from April to June.

Ericaceae—Heath Family

Arctostaphylos uva-ursi
kinnikinnick

A native trailing, woody evergreen shrub that forms low mats, kinnikinnick is often used in horticulture as a ground cover. The plant can root from branches, making propagation easy. Leathery leaves are oval to spoon shaped, sometimes with a notch on the distal end. Urn-shaped, light-pink flowers grow in terminal clusters. Berries are bright red and provide food for wildlife. It prefers well-drained, dry soil and full sun on slopes and forest edges from low to high elevations. It blooms from April to June.

Fabaceae [Leguminosae]—Pea Family

Acmispon parviflorus [Lotus micranthus]
short-flower bird's-foot-trefoil
small-flowered deer vetch

This annual native herb grows from the coast to the west side of the Cascades from British Columbia to California. Mostly glabrous, the thin stems grow from 10 to 30 cm long, either erect or prostrate, and are often branched, especially at the base. The leaves are odd-pinnate with 3 to 5 leaflets, oblong to oblong-obovate in shape and measuring 5 to 12 mm long. From April to September, the small pea-like flowers are borne singly and open creamy yellow tinged with red, often fading to pink or salmon. Fruits are 15 to 30 mm long legumes, that produce between the 4 to 8 seeds.

Cytisus scoparius
Scotch broom, Scot's broom

An invasive non-native shrub, Scotch broom grows up to 3 m tall with green ribbed branches. The leaves are trifoliate at branch bases and simple above. The individual flowers resemble pea blossoms, with the characteristic banner, wing and keel petals. They are usually solitary in leaf axils along the stems and deep yellow, sometimes tinged red or purple. The fruit resembles a flattened pea pod with hairs on the margins. At maturity, they burst open, flinging the seeds great distances where they can lay dormant for decades. Scotch broom grows rapidly in disturbed open habitats and has become a significant threat to our prairies. It blooms from April to June.

Lupinus albicaulis
sicklekeel lupine, white-stemmed lupine

This native perennial herb grows in open, well-drained sites. Its leaves are palmately compound and composed of 5 to 10 narrowly oblong leaflets measuring up to 7 cm. The stems and both upper and lower surfaces of the leaves are covered in fine whitish hairs, giving it one of its names. Sicklekeel lupine blooms from May to July. The bilaterally symmetrical flowers vary in color and can be white, yellow, purple, or blue and white. They are arranged in whorls on vertical racemes up to 44 cm long. Flowers have an upcurved keel and are usually 10 to 16 mm long. This pea family member produces silky-hairy pods up to 5 cm in length, with numerous seeds inside.

Lupinus bicolor
two-color lupine, miniature lupine, small-flower lupine

A native annual, two-color lupine grows to 40 cm and produces one flowering stem that is sometimes branched. Flowers are pale blue to deep purple with a white area of the banner, often having blue or purple spots, that turns magenta after pollination. The flowers are arranged in racemes. Flowering stalks and sepals are hairy, while the palmately compound leaves, composed of 5 to 7 leaflets, are sometimes hairy on the top surface. This relatively abundant species blooms from April to July.

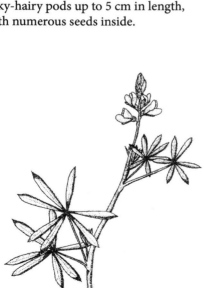

Lupinus lepidus var. *lepidus*
prairie lupine

A low and spreading native perennial herb, prairie lupine features racemes of blue flowers, rising above predominately basal palmately compound leaves covered with fine hairs. *Lupinus lepidus* is a species featuring several geographical variants, of which *L. lepidus* var. *lepidus* is found on south Puget Sound prairies. A montane variant of the species, *L. lepidus* var. *lobbii,* with reduced leaves and more compact inflorescences grows in rocky alpine habitats in the Cascades.

Lupinus polyphyllus var. *pallidipes*
bigleaf lupine, large-leaved lupine

Reaching 1.5 m tall, this native perennial lupine has stout, hollow stems rising from a woody rhizome. Leaves are palmately compound and composed of 10 to 17 leaflets, each up to 6 cm in length. Basal leaves feature long petioles, up to 6 times the length of the leaves themselves. Tight clusters of blue to purplish flowers 15 to 40 cm long add color to moist, open sites from June to August. Bearing 4 to 8 seeds each, the hairy pods of bigleaf lupine grow to 5 cm long.

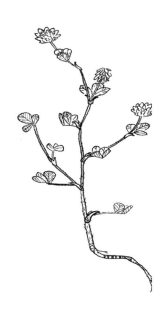

Trifolium dubium
least hop clover, suckling clover

An introduced annual herb, least hop clover is a species that prefers disturbed and grassy areas and is found abundantly in the South Sound prairies. Many small (3 to 3.5 mm), yellow flowers are clustered in heads less than 8 mm across and are not subtended by an involucre. Trifoliate leaves are somewhat hairy and leaflets can be up to 2 cm long. The upper half of the leaves have tiny teeth on their margins. This little clover blooms throughout the summer and produces fruits that are typical pea pods about 3 mm long.

Trifolium pratense
red clover

Red clover is a common forage crop in the Pacific Northwest. Introduced from Eurasia, it has become widely distributed through the U.S. It is a biennial or short-lived perennial herb, often soft and hairy, growing several stems from 30 to 100 cm tall. The leaves are trifoliate, with lanceolate to oblong-obovate leaflets that have prominent greenish veins and often exhibit a light green to white V-shaped marking. The flowers, which appear from June through August, are pink to purplish-red and borne in dense globose to conic-shape terminal heads. The fruits are 2-seeded legumes approximately 4 to 5 mm long.

Trifolium subterraneum
burrowing clover

Burrowing clover is a hairy non-native annual herb with prostrate to creeping stems that often root at the nodes. It inhabits pastures, prairies, roadsides and other open, disturbed areas at low elevations mostly west of the Cascades from British Columbia to California. The trifoliate leaves are borne alternately or in clusters and have wide, tapered stipules. Leaflets are egg to heart shaped, rounded and notched at the tip, 10 to 15 mm long, and often display light colored V-shaped markings. Blooming from April to June, 2 to 7 white or cream to pinkish pea-like fertile flowers are borne in terminal inflorescences on long stalks. In the center of the inflorescence, sterile flowers consisting of curved, rigid, branched calyces gradually develop above the fertile ones. In time, these become reflexed over the developing one-seeded pod, creating a burr-like structure. These structures tend to become buried, hence the plant's name.

Vicia americana var. *americana* [V. a.]
American vetch

This native herbaceous, climbing perennial likes open fields and open forests at low to middle elevations. Its pinnately compound leaves have 8 to 12 opposite leaflets and a single tendril at the tip. American vetch bears purple flowers that look like small pea flowers and are usually in inflorescences of 4 to 10. Fruits are hairless legumes that grow to 3 cm long. It blooms from May to July.

Vicia sativa var. *angustifolia*
common vetch

A non-native perennial herb with a climbing/creeping habit, common vetch grows mainly in disturbed areas at low elevations all over the U.S. Its leaves are pinnately compound with 10 to 15 leaflets and a winding tendril at the tip. Leaflets have a needle-like tip and tiny hairs lining the entire margins. Leaflet shape is linear to obovate-oblanceolate. Blooming from May to July, the flowers are pink-purple and pea-like, bilaterally symmetrical with a prominent banner. Fruits are legumes 3 to 7 cm long. As compared to *V. americana* var. *americana*, the flowers are darker in color, the banner larger and flatter, leaves narrower and the plant is smaller.

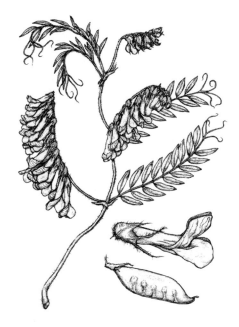

Vicia villosa var. *villosa*
[V. v. in part*]*
hairy vetch, woolly vetch, winter vetch

Hairy vetch is an annual or biennial herb that blooms from June to September. Climbing or sprawling, the stem reaches 0.5 to 2 meters in length, and is covered with long, spreading hairs. The leaves are pinnate with 8-12 pairs of linear-lanceolate leaflets and a well-developed, terminal tendril which clings to the surrounding vegetation. The inflorescence is a slender, 1-sided raceme arising from the leaf axils, each with 10-30 pairs of pink to blue-violet flowers. Each flower is about 15-18 mm long and pea-like, consisting of five petals fused into a long tube and a hairy calyx united at the base. The style is densely bearded at the tip. Seeds are born in flat-sided pods that can grow up to 10 mm long. Introduced from Europe, this vetch is distributed widely across most of North America where it thrives in disturbed meadows, prairies, roadsides and other open habitats. It is often an escapee from agricultural areas, where it is sometimes used for covercropping, and can form large, vegetative colonies in addition to reproducing sexually.

Fagaceae—Beech Family

Quercus garryana var. *garryana*
[*Q. g.* in part]
Oregon white oak, Garry oak

A stout limbed native deciduous tree reaching 25 m tall, Oregon white oak is found from southern British Columbia to central California. The trees produce unisexual, inconspicuous flowers before leafing out in the spring. The leaves of Oregon white oak feature rounded lobes and a dark green color. Oaks growing in large openings take on a savanna form, with a more or less circular outline, while woodland oaks have longer trunks and narrower crowns. These oaks are threatened by conifer encroachment into prairie habitats. The slower growing, shade-intolerant oaks will languish and eventually die once over-topped by conifers. The nutritious acorns of this oak have historically provided an abundance of food for Native peoples, who process them by soaking to remove the bitter, tannic flavor of the raw nut.

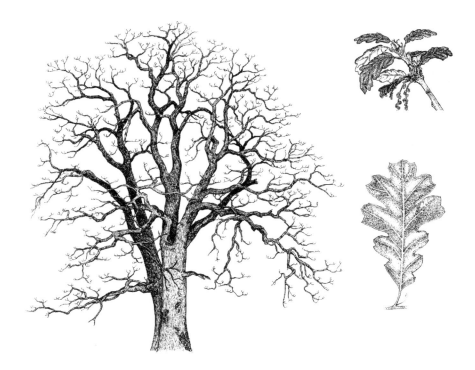

Geraniaceae—Geranium Family

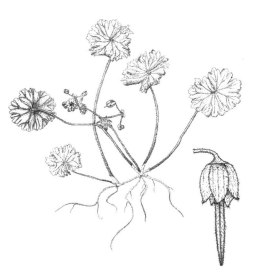

Geranium molle
dovefoot geranium

This introduced annual herb has hirsute stems, leaves, and sepals. The tap-rooted herb is 10 cm to 40 cm tall. Reniform, or kidney-shaped, leaves are divided into 5 to 7 lobes that have round teeth with slightly pointed tips. Its pink flowers usually bloom in pairs, with five bi-lobed petals. The fruit looks like a "beak," hence the common name, "crane's-bill" for the genus. Thrives in moist disturbed areas at low elevations. Native to Europe, it blooms April to September.

Hypericaceae—St. John's-wort Family

Hypericum perforatum
common St. John's-wort

A perennial herb with a taproot and short rhizomes, common St. John's-wort grows up to 1 m tall. The lance-shaped to obovate leaves have punctate glands and often a purple or black tint. Common St. John's-wort has branching inflorescences of flowers with 5 yellow petals and 5 sepals. Well known for its medicinal properties, this herb is a non-native introduction from Europe. It grows in pastures and disturbed areas and is especially common throughout the west coast of North America. In some areas it is classified as a noxious weed. It blooms in June and July.

Lamiaceae (Labiatae)—Mint Family

Clinopodium douglasii
[Satureja douglasii]
yerba buena

Occasional in prairies adjacent to woodlands and forests, yerba buena has a prostrate habit with stems running along the ground. Aromatic leaves are egg-shaped and feature softly serrated margins. White to purplish flowers are borne from short stalks growing from leaf axils. Flowers are bilabiate, with a spreading 3-lobed lower lip and a 2-lobed upper lip.

Lamium purpureum
dead-nettle

Dead-nettle is a non-native annual herb with a short taproot. A member of the mint family, it strongly resembles culinary mint in its leaf structure and growth habit. Square-shaped stems grow up to 40 cm tall. Leaves are pubescent, more or less heart shaped, with the upper leaves particularly having a purple tinge to them. Some plants are intensely purple-red. Flowers are pale pink-purple, bilabiate, and have a ring of hairs at the base of the corolla tube. It is common in disturbed areas at low to mid-elevations, and blooms from April to July.

Prunella vulgaris
self-heal, heal-all

Self-heal is a perennial rhizomatous herb growing either erect or decumbent, with either solitary or clustered stems from 10 to 50 cm tall. The flowers are usually blue-violet and borne in dense bracteate terminal spikes that some children call "bumble-bee flowers." As typical of the mint family, it has square stems, but it is not noticeably aromatic. The elliptic to broadly ovate leaves with entire margins are opposite on the stem as well as basal. Self-heal is commonly found in moist areas from sea level to mid-elevations, blooming from May to September. Two varieties can be found growing across North America: *P. vulgaris* var. *lanceolata* is the native form and *P. vulgaris* var. *vulgaris* is introduced from Europe. As its common name suggests, this herb has a long history of use in healing on both continents.

Onagraceae—Evening Primrose Family

Chamerion angustifolium [*Epilobium angustifolium*]
fireweed

This common native perennial herb sports a long showy inflorescence of deep purplish-pink flowers at the apex of each stem. Flowers consist of 4 linear sepals alternating with 4 rounded petals. Stigmas are exerted, with 4 clefts. The leaves are narrowly lanceolate with entire margins and arranged alternately along the stem. It spreads by rhizomes, forming large showy patches, particularly in burned areas. In late summer to fall, the linear pods open to release tiny seeds with silky hairs that enable them to disperse on the wind. A circumboreal species, fireweed grows 1 to 3 m tall in open areas from the coast to the subalpine zone in the mountains. It blooms from June to September.

Clarkia amoena var. *lindleyi*
farewell-to-spring

Farewell-to-spring is a native annual herb with flowers that close in the nighttime. Leaves are simple and alternate. Flowers have 4 pink to pale purple petals, 1 to 4 cm long, sometimes with a reddish spot in the center of each. The eight stamens are of two different lengths. The style parts to a 4-lobed stigma, each 1.5 to 5 mm long. Ranging from British Columbia to California, west of the Cascades and in the Columbia River Gorge, farewell-to-spring blooms from May to July.

Orobanchaceae—Broomrape Family

Orobanche uniflora
naked broomrape

This parasitic native annual plant grows in dry, generally open areas on both sides of the Cascades in Washington and is broadly distributed throughout North America. Plants feature 1 to 3 flowers borne singly on leafless stalks to 5 cm long, generally appearing from April to August. The 5-lobed flowers form a tube and are purplish with a yellow throat. Unlike chlorophyll- containing green plants, broomrape is unable to produce food via photosynthesis. Instead, it parasitizes photosynthates from host plants via specialized roots called haustoria. Though some members of the genus *Orobanche* require a specific host plant, naked broomrape parasitizes a fairly broad variety of plant genera.

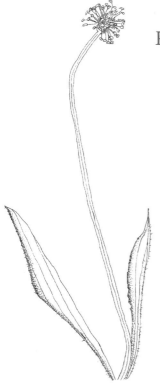

Plantaginaceae—Plantain Family

Plantago lanceolata
narrowleaf plantain, English plantain

Narrowleaf plantain is a non-native perennial herb that grows from 15 to 45 cm tall. It has a fibrous root system and underdeveloped taproot. The basal leaves have parallel veins, are elliptic to lance shaped, and 10 to 40 cm in length. The flower stalk is long, with 5 longitudinal lines or grooves. The terminal inflorescence or spike is green, and oval-shaped at the bud stage. When in bloom, it has whorled, inconspicuous, white flowers with prominent stamens. The seeds are long, shiny and black. Narrowleaf plantain is a common weed found in pastures, roadsides, and other disturbed sites. It is native to Eurasia and has become a cosmopolitan weed, often found in moist areas. It blooms from April to August.

Plantago major
common plantain, great plantain, nippleseed

Common plantain is a perennial herb introduced from Europe that has become a cosmopolitan weed distributed widely throughout North America. A basal rosette of broadly ovate leaves that contract abruptly to the petiole grows from a short, stout woody base. Leaves are 4 to 18 cm long, have entire or irregularly toothed margins and are often deeply ridged or crinkled. Small non-showy flowers are borne from April to August on a dense narrow spike 5 to 30 cm tall. The fruits are tiny capsules 2.5 to 4 mm long. Found on roadsides, in ditches and other disturbed areas, common plantain is another species that was brought by European settlers as a medicinal herb.

Plantago patagonica
hairy plantain, Indian wheat

Common east of the Cascade Range, hairy plantain is an extremely rare species on the South Sound prairies. Its English name is a reference to the dense, silvery hairs on its leaves and flowering stalk. This species of plantain is an annual growing to 15 cm tall with narrow, wholly basal leaves lacking petioles. Flowers are clustered on woolly spikes, which generally grow taller than the length of the leaves. Individual flowers are more conspicuous than in our other plantain species with 4 white to translucent petals, 4 stamens, and a feathery stigma adapted for catching pollen on the wind. Hairy plantain blooms from April to June.

Plumbaginaceae—Plumbago Family

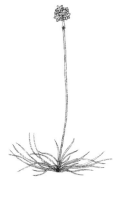

Armeria maritima ssp. *californica*
thrift, sea-pink

Thrift is a native perennial herb with pale green, densely bunching, persistent, linear basal leaves. Flowering heads are borne on an unbranched scape 10 to 50 cm tall. Flowers are pink, having 5 petals connate at the base and occur in dense clusters. The calyx is dry, papery and funnel-shaped. It is generally a coastal plant, as suggested by one of its English names, but it is occasionally found in inland prairies. This charming little plant is circumboreal in distribution and blooms from March to July.

Polygonaceae—Buckwheat Family

Rumex acetosella
sheep sorrel

Sheep sorrel is an herbaceous plant that grows up to 50 cm tall. It is an introduced species that is rich in vitamin C and has a pleasant, sour taste. Sheep sorrel is dioecious, with male and female flowers on separate plants. Red or yellowish flowers are borne on upright panicles. The leaves have sheathing stipules and are highly variable. They can be linear to ovate, though usually arrow-shaped and 2 to 4 cm long with triangular lobes. Basal leaves are numerous, cauline leaves less so. Sheep sorrel can be found in disturbed sites at lower elevation, especially roadsides and fields. It blooms from May to August.

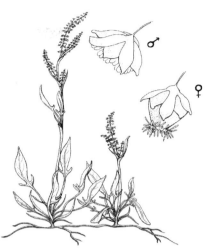

Portulacaceae—Purslane Family

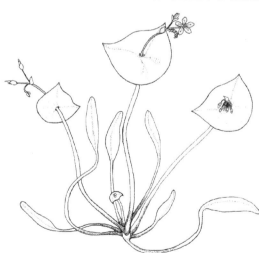

Claytonia perfoliata
[Montia p. in part]
miner's lettuce

Miner's lettuce is a fleshy native annual or perennial herb, 1 to 5 cm tall, growing from a taproot but often developing rhizomes. This herb has elliptical basal leaves 1 to 4 cm long that are sometimes tinged purple or brown. Below an inflorescence of 5 to 40 flowers, a pair of leaves are fused around the stem to form a shallow cup. The flowers are small, white to dark pink in color with dark veins. They have 5 petals with a notch at the tip and lavender stamens. This edible plant can be found in moist shaded areas, common under oak trees. It ranges from Alaska to southern California on both sides of the Cascades and eastward toward Montana and Utah. It blooms from March to July.

Primulaceae—Primrose Family

Dodecatheon hendersonii
Henderson's shooting star, broad-leaved shooting star

Henderson's shooting star is a native prairie perennial whose flowers emerge from March to June. Deep magenta flowers point towards the ground and 5 petals are peeled back to create an inside-out look. There is a yellow and white ring around the base of the petals. A deep red to purple tip composed of stamens surrounds a single pistil. Flowers are borne in umbels on single scapes, 10 to 30 cm tall. Thick, fleshy leaves are usually glabrous and form basal rosettes. The leaves have a lighter colored center vein and entire margins.

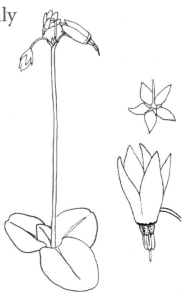

Dodecatheon pulchellum var. *pulchellum*
dark-throated shooting star

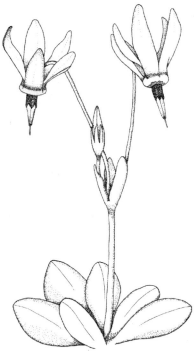

This beautiful native perennial herb blooms April to July. The leaves are basal, 2 to 15 cm long and narrow with winged petioles, and smooth and entire with a visible center vein. Flowers are borne atop 5 to 40 cm stems in involucrate umbels, each turned inside out and downward with 5 pink-purple petals. Stamens are clustered around the style with a fused yellow, orange or purple filament tube and red, dark purple or yellow anthers. The coloring of the reproductive organs is highly variable and probably dependent on the age of the flower and the environment. Very similar in appearance to *Dodecatheon hendersonii* (above), but less common in the prairies. There are some key differences in appearance, especially in the leaves. Leaves are thinner, lighter in color, and narrower, with more prominent petioles than *D. hendersonii*.

Ranunculaceae—Buttercup Family

Anemone lyallii
Lyall's anemone

Lyall's anemone is a native perennial herb featuring a single flower with 5 showy white (seldom pale blue or pink) sepals and no petals. Leaves are composed of 3 leaflets with toothed margins. Found in moist prairies and forest edge habitats, Lyall's anemone is named for Scottish surgeon David Lyall (1817-1895), who collected the plant for introduction into English and Scottish gardens. It grows 5 to 25 cm tall and blooms from March to July.

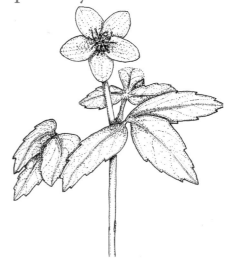

Aquilegia formosa var. *formosa*
western columbine

This common perennial herb grows from a taproot and produces an erect stem up to 1 m tall each with 2 to 5 flowers, sometimes more in vigorous plants. The stems are hairless below but sparsely hairy and somewhat glandular in the inflorescence. The mostly basal leaves are twice divided in 3s with blades that are green above and glaucous beneath. Striking, pendent, mostly red flowers sport some yellow and have showy, exerted stamens. Hummingbirds relish the nectar from the 5 long spurs with bulbous tips. The fruits are 5 or more erect follicles with hairy, spreading tips and numerous black, wrinkled seeds. Western columbine prefers moist sites that are open or partly shaded from the lowlands to timberline in the western United States and western Canada. It blooms from June to July.

Delphinium nuttallianum
upland larkspur

This showy native perennial herb has flowers borne on 14 to 40 cm stems in racemes of 3 to 15. The 5 sepals, widely spreading, are the showy part and are variable in color, often intense purple-blue. They measure 17 to 25 mm long with the lower pair being the longest. The 13 to 20 mm spur is also richly colored and is from about as long to twice as long as the top sepal. The petals are small, blue or sometimes pale blue with dark veining. The leaves are palmate and deeply lobed, mostly basal with cauline leaves greatly reduced. It blooms from March to July.

Ranunculus occidentalis var. *occidentalis*
western buttercup

This native perennial is more mild-mannered than its relative, *R. repens*, which takes over gardens and pastures with invasive glory. *R. occidentalis* is found as a prairie wildflower. The petals are wider and leaf serrations less pointed than *R. uncinatus* (below). Growing 15 to 40 cm tall, it blooms April to June right before and among the camas. It bears 5-petaled yellow flowers with many stamens and pistils atop numerous erect pubescent stems with 3-lobed serrate leaves. The fruit is an aggregate of pumpkin seed-shaped achenes. The styles jutting out from the ends of the achenes are straight, not hooked like *R. uncinatus*. Leaves are more pubescent and lighter in color than *R. uncinatus*.

Ranunculus uncinatus
little buttercup

This hairy native perennial herb grows 20 to 60 cm tall and has 3-lobed, serrate lower leaves that become lanceolate towards top of plant. Leaves are more deeply lobed and pointed than *R. occidentalis*. Flowers have 5 small, pale yellow petals, each with a small nectary at base. The aggregate fruit is made up of multiple achenes, each bearing a tiny hook visible without magnification. In general, it is less hairy than *R. occidentalis*. It grows along roadsides and forest edges and blooms from April to July.

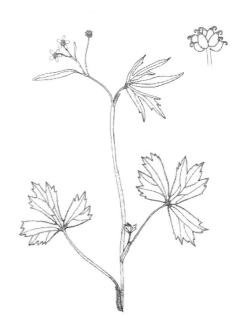

Rhamnaceae—Buckthorn Family

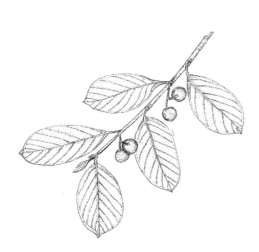

Frangula purshiana
[Rhamnus purshiana]
cascara sagrada

A small tree generally up to 10 m tall, cascara sagrada typically grows in lightly wooded habitats but can be found in open areas. Leaves are prominently veined, oppositely arranged, and more or less egg-shaped. During the winter the buds are conspicuously lacking the typical protective bud scale employed by many deciduous species. Small greenish-yellow flowers appear between April and June. They develop into black berries by late summer. The "sacred" bark of this tree has been long-used as a laxative (stimulant).

Rosaceae—Rose Family

Amelanchier alnifolia
serviceberry

Serviceberry is a native shrub or small tree, usually growing 1 to 2 m tall. Young branches are a reddish brown and smooth, eventually fading to gray. The leaves grow on thin petioles with oval to oblong shape and are usually broad with fine serration at the tips. The flowers are white with narrow petals. Round, dark purplish fruits are juicy and traditionally have been an important food source for Native peoples. Blooms from April to July.

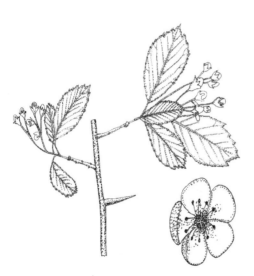

Crataegus douglasii
black hawthorn

This thorny, native deciduous shrub grows up to 10 m tall. Leathery leaves are weakly lobed at the distal end and pale on the undersides. White flowers with a free hypanthium bloom in clusters at the terminal end of branches or in the leaf axils. The pomme fruits ripen to a purplish black. Black hawthorn prefers open spaces at habitat borders, such as stream edges, roadsides, forest edges, and shorelines in low to mid-elevations. It blooms from May to June.

Crataegus monogyna
one-seed hawthorn

A naturalized European shrub or small tree, this species grows from 2 to 10 m tall. The leaves are deeply lobed, splitting the leaf to half its width, with 3 to 7 lobes. White, red or yellow flowers, produced from May to June, result in only 1 or 2 seeds compared with many in *C. douglasii*. One-seed hawthorn also prefers open spaces at habitat borders, such as stream edges, roadsides, forest edges, and shorelines in low to mid-elevations.

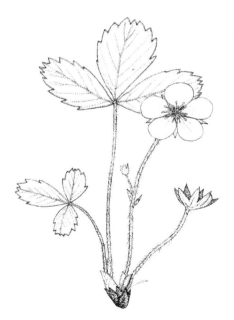

Fragaria virginiana
wild strawberry

This native perennial herb likes open areas in the woods, plains or gravelly meadows. The leaves are trifoliate with the petioles up to 15 cm. The leaf margins are toothed mostly at the tips. Leaves can have a bluish tinge, and are slightly frosted looking from small white hairs on the leaf surface. Stems are pubescent and red, as is the calyx. Blooming from May to August, the flowers have 5 rounded white to pinkish petals. The fruit is really an enlarged receptacle with multiple pistils on the surface which, in maturity, look like yellow seeds. This sweet, red fleshy "berry" is highly prized by both animals and humans.

Oemleria cerasiformis
Indian plum, osoberry

Indian plum is a native deciduous shrub with alternate, light green, lanceolate leaves. It usually grows between 1.5 to 3 m tall but sometimes can reach up to 5 m. A very early bloomer, it produces a pendent raceme of white, bell-shaped flowers in early March to April. The plant is dioecious–male and female flowers are on separate individuals. The edible fruits are like miniature plums that age from cream-colored with a blush to dark, bluish black. Indian plum is found at low elevations, on roadsides, open areas, stream banks, or woods with dry to moist soil.

Potentilla gracilis var. *gracilis*
slender cinquefoil

Slender cinquefoil is a native perennial herb that is highly variable. It grows several erect stems 40 to 80 cm tall from a branched crown with alternately arranged, palmately compound leaves. The leaves are white woolly on the underside, display large stipules, and are comprised of 5 to 9 leaflets, oblong-eliptic in shape with deeply toothed margins. The 5-petaled flowers are yellow and occur in somewhat flat-topped open clusters. It occurs in varied habitats throughout Washington and blooms from July to August.

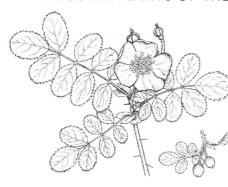

Rosa gymnocarpa
bald-hip rose

This native perennial shrub inhabits woodlands, meadows and prairies. Though a sun-lover like all roses, it is more shade tolerant than most, and can often be seen growing happily in oak or alder groves, where it sometimes forms small thickets. It has an upright growth habit with straight slender stems up to 2 m tall. The prickles are straight, fine, sharp and needle-like and vary considerably in density. Some canes are so bristly they resemble brushes, while others are nearly prickle-free. The pinnately compound leaves are composed of 5 to 9 elliptical light green leaflets with serrated edges. May through August this charming rose is adorned with small, often intensely colored 5-petaled blossoms. Measuring 3 to 4 cm across, they range in color from light pink to deep rose with bright yellow stamens. The 1 to 1.5 cm oval fruits are bright red and are often retained through the winter. They are unique in that they shed their calyces, becoming "bald" and resembling berries.

Rubus spectabilis
salmonberry

This native perennial shrub blooms March to June and bears edible, raspberry-like aggregate fruits that are yellow to red in color. Its habit and overall appearance is very similar to the cultivated raspberry, growing to a height of from 1 to 3 m. New shoots have reddish skinned stems, while old stems have brown shedding bark, typically mildly armed with short prickles. Leaves are trifoliate with leaflets that are roughly ovate and doubly serrate, and almost hairless. Flowers each have 5 bright magenta petals that appear crinkled, 5 sepals, and many pistils and stamens.

Rubus ursinus
trailing blackberry, dewberry

A native trailing perennial vine/shrub with erect floral branches, trailing blackberry has flowers with 5 distinct white petals. Male and female flowers are borne on separate plants. Trifoliate alternate leaves are egg-shaped to lance-shaped with serrated edges with the terminal leaflet usually longer than the others. Long, trailing stems armed with abundant, slender, hooked prickles grow up to 6 m long. The highly prized edible fruit is composed of black to dark purple druplets. Trailing blackberry can be found in prairies, open to dense woodlands, and in disturbed sites. It ranges from British Columbia to northern California and eastward into central Idaho, blooming from April to early August.

Rubiaceae—Madder Family

Galium aparine
cleavers, bedstraw, goosegrass

This native annual herb grows up to 1 m tall with weak stems. It tends to grow upon other plants for support, using the tiny hooked hairs that cover the plant to grip. The leaves are slender and occur in whorls. The inflorescences are small and borne on peduncles in leaf axils, with 4 white petals and 4 sepals. The fruit is segmented into 2 spheres covered in hooked bristles, which help disperse the seed on the coats of animals. This common weedy species is found in a variety of habitats in temperate North America. It blooms from April to June.

Galium parisiense
wall bedstraw

A weedy European species, wall bedstraw is a branching, tap-rooted annual with many tiny (1 mm), white flowers and square stems. Leaves are generally in groups of 5 to 8 and whorled at the nodes. The plant is herbaceous and covered in tiny, hooked hairs, which make it feel sticky. Flowers have 4 to 5 petals and arise from the axils of the leaves in small, open clusters. The indehiscent 2-parted fruit is easily recognizable. Wall bedstraw blooms from April through August.

Sherardia arvensis
blue field madder

Originally from the Mediterranean, blue field-madder is a small annual herb that somewhat resembles wall bedstraw. It is a fairly common weed in open areas in the far western United States. Leaves grow in whorls of 6 and have stiff hairs on the upper surface. Flowers arise on short peduncles from the axils of the leaves and are each subtended by a toothy involucre. Flowers are very tiny at about 3 mm, pinkish to purplish, and bloom from April to July.

Salicaceae—Willow Family

Salix scouleriana
Scouler's willow

A shrub or small, multi-trunked tree to 15 m in height, Scouler's willow grows in moist sites from Alaska to California on both sides of the Cascade Range. Younger stem growth is generally yellowish to brown in color, while older stems and the trunk are typically light gray. Leaves are alternate, mostly oblanceolate, with top surfaces somewhat glossy and the bottom surface generally glaucous, often featuring fine, matted hairs. Mature leaves lack stipules. Unisexual flowers in dense catkins generally appear from March to June before the leaves, with male catkins 2 to 4 cm in length and female catkins slightly longer. Willows can be quite difficult to identify to the species, and the occasional presence of hybrids can make things worse. In the case of Scouler's willow, identification can be somewhat aided by habitat, as it is one of a few willows that will grow in upland sites west of the Cascades, away from the riparian areas that typify willow habitat in the Pacific Northwest.

Saxifragaceae—Saxifrage Family

Lithophragma parviflorum
small-flower woodland star, small-flower prairie star

This delicate wildflower sports a glandular-pubescent stalk that ranges from 10 to 30 cm in height. The perennial herb grows from slender rhizomes that produce rice-like bulblets. Most of the leaves occur basally with blades that are usually divided to the base into 5-lobes that are each further cleft. The lobe segments become longer and narrower in the cauline or stem leaves. Its inflorescence is a raceme with 5 to 11 white flowers each with 5 deeply tri-lobed petals. Each petal narrows dramatically to its insertion point, a characteristic also known as clawed. Small-flower prairie star blooms March to May in oak woodlands and along their grassland interface.

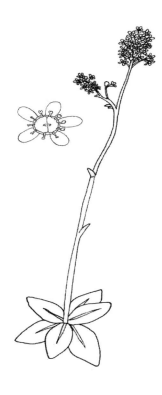

Micranthes integrifolia [*Saxifraga integrifolia*]
swamp saxifrage, whole-leaf saxifrage

Swamp saxifrage is a perennial native flower that prefers the moister areas in the prairie, such as near a stream or tree. Leaves form a basal rosette. Leaves have smooth margins, a visible center vein, are 2 to 5 cm long and are very rounded. They can seem thick enough to be succulent. Panicles of small flowers are borne on hardy, pubescent stems 10 to 30 cm tall that become increasingly red toward the terminal inflorescence. The 1.5 to 3 mm flowers have 2 obviously separate pistils and a perigynous disk with 10 stamens and 5 round, spaced, white petals attached on the edge. Anthers are yellow, ovaries are green and can be red as fruit matures. It flowers late March through July.

Tellima grandiflora
fringe cup

Fringe cup is a very hairy rhizomatous native perennial herb with shallowly lobed, basal and cauline leaves. Leaves are oval to heart-shaped and doubly-serrate, decreasing in size higher up the plant. The fragrant flowers, borne in narrow racemes up to 80 cm tall, are complete, with small, reflexed, pinnately divided, greenish-white to reddish petals. It is commonly found in moist forests, riparian zones, and clearings, blooming from April to July.

Scrophulariaceae—Figwort Family

Castilleja hispida var. *hispida*
harsh paintbrush

Harsh paintbrush is a woody-based native perennial herb with alternate, lobed leaves. The flowers feature a corolla tube surrounded by leafy bracts, gradating from green basally to red at the tips. The plant is mostly covered with fine hairs. Like many other members of the *Castilleja* genus, our paintbrush species are considered hemiparasitic, with the plant augmenting its nutrient uptake by connecting to the roots of other plants. It grows 20 to 60 cm tall and blooms from late April to August.

Castilleja levisecta
golden paintbrush

Like the harsh paintbrush, golden paintbrush is a perennial hemiparasitic herb. Alternate leaves and bracts are oval and lobed in pairs of 2 to 4. The bracts are bright yellow and each subtends a flower. Flowers are complete, bilabiate with lower petals very reduced. Stems and leaves are very pubescent. Non-branching, it reaches approximately 10 to 50 cm in height, growing in small, but dense clusters. A federally-listed threatened species, it blooms from April to September.

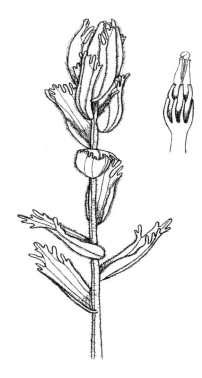

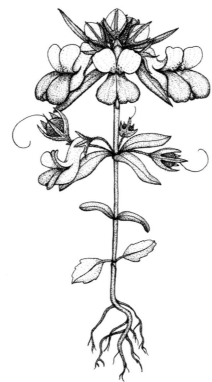

Collinsia grandiflora
large-flowered blue-eyed Mary, giant blue-eyed Mary, blue-lips blue-eyed Mary

This native annual produces erect stems that are simple or branched. The whole plant ranges in height from 5 to 40 cm and is minutely glandular-hairy throughout, with opposite leaves. Lower leaves are long-stalked, often toothed, and relatively broad, and become increasingly unstalked and narrowly linear as they move up the stem. Flowers are strongly 2-lipped, with a deep blue to violet lower lip and a paler upper lip. The bilaterally symmetrical flowers occur in the leaf axils generally occurring singularly at the lower nodes and clustered at the terminal node. Each bicolored flower is borne on a short, hairy stalk. The calyx is 5 to 8 mm long and 5-lobed, out of which the tubular corolla emerges at a sharp right angle. Though they have been known to hybridize, *Collinsia grandiflora* can be distinguished from its close relative, *Collinsia parviflora*, by its larger, more showy flowers (9 to 17 mm long) and its stouter, more erect habit. Large-flowered blue-eyed Mary blooms from April to June in open, vernally moist to rather dry, grassy areas.

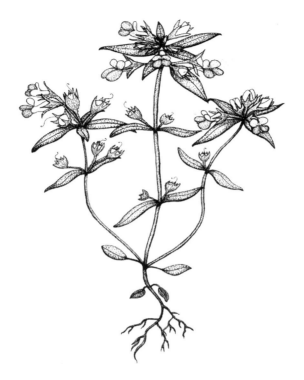

Collinsia parviflora
small-flowered blue-eyed Mary, maiden blue-eyed Mary

This herbaceous native annual is often found growing in masses with minutely hairy stems that are ascending to erect and sometimes sprawling, reaching 5 to 50 cm in height. Its leaves are opposite, smooth to minutely hairy, and often purplish beneath. Lower leaves are egg- to spoon-shaped on long stalks. The middle and upper leaves are oblong to linear-lanceolate and 1 to 4 cm long. They become bract-like, smaller, linear and often whorled as they approach the terminal inflorescence. The flowers of *Collinsia parviflora* are 2-lipped, with a blue lower lip and pale pink to white upper lip, and are much smaller than those of its close relative, *Collinsia grandiflora* (4 to 7 mm long). It can also be distinguished from the large-flowered blue-eyed Mary by the oblique angle of its bent corolla as viewed from the side. The small-flowered blue-eyed Mary is more widely distributed and as a result blooms over a longer period of time than the large-flowered blue-eyed Mary. Its flowering time spans from March-July beginning in the lowlands on both sides of the Cascades and ending in alpine meadows.

Scrophulariaceae

Nuttallanthus texanus
[Linaria canadensis var. texana]
blue toadflax

This native annual or biennial forb grows upright from 10 to 60 cm tall, with decumbent side shoots that do not flower. The narrow, obtuse cauline and basal leaves are generally glabrous, but the infloresence is sticky-hairy (glandular puberulent). The blue to violet flowers are borne alternately in a raceme. Each blossom is bilabiate with a smaller upper lip and a 3-lobed lower lip that sports a slender curved nectar spur. The ridge in the center of the lower lip that leads into the corolla throat is white. This herbaceous plant is found in disturbed areas with sandy or gravelly substrate below 1800 m elevation. The range of blue toadflax extends from prairies in the San Juan Islands south to California. It blooms from April to May.

Parentucellia viscosa
yellow glandweed

This weedy annual herb, introduced from the Mediterranean, grows upright and unbranched from 10 to 70 cm tall. The leaves are strongly toothed, sticky-hairy (glandular pubescent), opposite, and sessile, with no basal leaves present. The yellow flowers are borne alternately in a terminal spikelike raceme. They are bilabiate in form, with a hood-like upper lip and a fuzzy, 3-lobed lower lip. The 4 stamens occur in 2 sets of 2 different lengths. It is found in moist areas in low elevations west of the Cascades from British Columbia to California, blooming from June to August.

Triphysaria pusilla
[Orthocarpus pusillus]
dwarf owl-clover

This native annual plant could easily go unnoticed with its slender form and small flowers. Leaves feature narrow lobes, and along with the stem are coarsely pubescent. Bilabiate flowers are reddish and 4 to 6 mm long, borne from axils of cauline leaves. Dwarf owl-clover grows in seasonally moist, open areas at low-elevations west of the Cascades.

Veronica arvensis
corn speedwell

This small taprooted annual herb was introduced from Eurasia and has spread across much of North America where it is found in disturbed ground, gardens, and on roadsides. Growing 5 to 30 cm in height, the somewhat hairy stems can be either upright or decumbent, simple or branched below. The slightly hairy bright green leaves are opposite, ovate to broadly elliptic with rounded serrations, 1.5 to 2.5 cm long and half as wide to nearly as wide as their length. The lower leaves have short petioles while the upper leaves are sessile. The blue-violet flowers appear from April to September in a terminal bracteate raceme with alternate bracts that are narrower than the leaves, each subtending a single flower. The flowers are quite small, measuring 2 to 2.5 mm wide and are 4-lobed with the upper lobe being the largest.

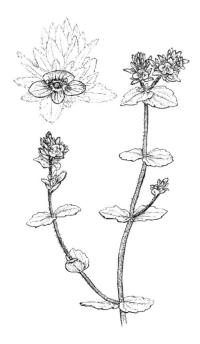

Urticaceae—Nettle Family

Urtica dioica ssp. *gracilis*
stinging nettle

This native perennial herb grows 1 to 3 m tall and spreads by rhizomes. The leaves are opposite and are lanceolate to broadly ovate with serrated margins. The leaves are armed with stinging hairs that contain formic acid, which causes skin irritation. The flowers are green in color and without petals, borne in clusters at leaf axils. The stinging nettle grows in a variety of habitats, from the sagebrush steppes along streams to moist and shaded lowlands. It is widespread throughout North America and Eurasia, and has long been an important source of food, medicine and fiber for many peoples. It blooms from May to September.

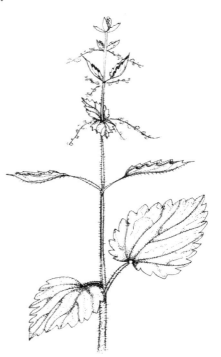

Valerianaceae—Valerian Family

Plectritis congesta
seablush, rosy plectritis

This native annual herb grows 10 to 60 cm tall. The leaves are opposite, oblong in shape, but more egg-shaped towards the bottom. The flowers are pink to white, have 5 petals and form in terminal clusters. This annual grows in moist meadows and on bluffs and rocky slopes often near the ocean, hence its common English name. Seablush can be found in western Washington and Oregon where it blooms from April to June.

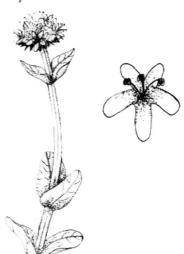

Violaceae—Violet Family

Viola adunca ssp. *adunca*
early blue violet

A native herbaceous perennial that grows up to 10 cm tall, early blue violet has ovate to cordate leaves. The blue to deep violet flowers are 5-petaled and sepaled and zygomorphic (irregular in symmetry). It thrives in disturbed areas, moist grasslands, meadows, and open forests at low to timberline elevations, blooming from April to August. *V. adunca* can be distinguished from the similar, low-elevation species, *V. howellii* by its long, narrow spurs and non-ciliate sepals. *V. howellii* has broad spurs and ciliate sepals.

Viola glabella
stream violet

Stream violet is a native perennial herb with entire, serrate, heart-shaped leaves with defined points at the tips. The flower stems reach 5 to 30 cm tall with no leaves on the lower two-thirds. They are yellow, 5-petaled and sepaled, complete, and zygomorphic. The lower 3 petals have purple penciling. It prefers moist areas, hence its common name. It blooms from March to July.

Viola praemorsa var. *praemorsa* [*V. nuttallii* var. *p.*], upland yellow violet

This perennial native violet blooms from April to July and is found primarily on prairies in Thurston County. Yellow flowers are 5-petaled and 5-sepaled, zygomorphic with 2 upper petals and 3 lower petals borne on stems of up to 15 cm. The bottom center petal is penciled with brownish-purple. The reproductive parts are hidden from plain view in a small chamber. Leaves are 3 to 10 cm long with 5 to 15 cm petioles. Leaf blades are pubescent, entire margined and teardrop shaped. Leaves have indented veins that almost appear parallel but are actually pinnately netted. The whole plant is short and dense, with flower stems originating basally.

Cyperaceae—Sedge Family

Carex inops ssp. *inops*
long-stolon sedge

A rhizomatous, though somewhat tufted plant, long-stolon sedge is a grass-like species growing in dry prairies and open woods west of the Cascades. Leaf blades are 1 to 3 mm wide and flat, typically shorter than the flowering stems. The solitary male (staminate) spike grows above several slightly congested female (pistillate) spikes, the lowest of which is subtended by a leafy bract 1 to 3 cm long. The fruit, called a perigynia, is a 3-sided achene surrounded by a special enclosed bract.

Iridaceae—Iris Family

Iris tenax
Oregon iris, tough-leaf iris

Oregon or tough-leaf iris grows in prairies and open forests west of the Cascade Range from Washington to northern California. Reaching 30 cm in height, the plant forms clumps from branched rhizomes. Leaves of Oregon iris are narrow and lanceolate. The basal leaves are lax and reach 45 cm, while leaves borne from the flowering stalk are generally up to 15 cm long and are more or less erect, sheathing the stalk for half their length. Flowers generally appear between April and June, ranging in color from the typical purple to yellow, or occasionally white. The unique morphology of irises requires some specific terminology to describe flowering parts: 3 erect petals (known as "standards") are subtended by 3 lax sepals (known as "falls"). The sepals feature a patch of yellow surrounded by a ring and pronounced venation (collectively known as the "signal"). The yellow patch on the sepals is an attractant to pollinating insects, especially bees. An oblong, 3-sided seed capsule 3 to 5 cm long forms after the flower has withered.

Sisyrinchium idahoense
blue-eyed grass

This native perennial grows 15 to 40 cm tall typically in wetter areas. It has basal, grass-like leaves with parallel venation. Blue-eyed grass has 6-tepaled purple-blue flowers with yellow centers. Tepals are lighter colored on the underside and have 3 to 4 distal visible parallel veins and a narrowed tip jutting out from the end. The flower has 1 prominent pistil and 3 stamens. The ovary is inferior and pubescent. Buds emerge in umbels of 2 to 5 from long narrow sheaths, much like irises. It blooms from March to June. Both *S. i.* var. *occidentale* and *S. i.* var. *segetum* have been reported from the south Puget Sound region.

Juncaceae—Rush Family

Juncus bufonius var. *bufonius* [*J. b.*]
toad rush

An annual native plant, toad rush is widely distributed in seasonally moist disturbed habitats. Narrow leaves, about 1 mm in width are exceeded by flowering stems 5 to 20 cm long. Green to brownish flowers occur singly at the nodes and feature a whorl of acuminate tepals surrounding and exceeding a 2 to 4 mm 3-celled capsule.

Luzula comosa var. *laxa*
[no Hitchcock and Cronquist synonym]
Pacific woodrush

Pacific woodrush is a native tufted perennial featuring reddish to green grass-like leaves with long fine hairs along the leaf edge. Stems reach 10 to 40 cm in height and bear an inflorescence of 1 to 6 brownish glomerules (a condensed, headlike cluster of flowers). Members of the rush family are differentiated from grasses in that they typically feature round stems without nodes. Also, they have 6 brown membranous tepals and thus very different floral morphology compared to grasses. This species blooms from April to July.

Liliaceae—Lily Family

Brodiaea coronaria
crown brodiaea, harvest brodiaea

Crown brodiaea is a native perennial herb with slender basal leaves that wither before the flowers bloom. Flowers usually occur in open clusters of several per plant with pedicels unequal in length. Each flower has 6 purple tepals joined at base. Reproductive parts include 3 fertile stamens and 3 infertile, flattened staminoidia. This lovely little flower, only 10 to 25 cm tall, prefers dry sites on grassy slopes, where it blooms in June and July.

Camassia leichtlinii ssp. *suksdorfii*
large camas

Large camas grows up to 30 cm tall from a single bulb. The long and slender leaves extend from the base of the plant. The regular flowers cluster in racemes 10 to 20 cm long and range in color from a dark violet to light blue. After the pollen is shed, the 6 tepals twist around the yellow-green ovary. Large camas favors moist meadows and prairies, blooming in April and May.

Camassia quamash
common camas

Growing from a single egg-shaped bulb, common camas is a native perennial that ranges from 30 to 70 cm tall. Flowers are borne in racemes and have 6 narrow tepals, light to dark blue in color. Close inspection of the flowers of camas show that they are slightly irregular-the lowest tepal curves upward away from the stem. The grass-like leaves are basal and exhibit parallel venation. Common camas grows in seasonally moist, open sites, blooming from late April through June. A major traditional food source for Native peoples, camas is processed for both immediate consumption and for long term storage. Both *C. q.* ssp. *azurea* and *C. q.* ssp. *maxima* are reported to occur in the prairies of the south Puget Sound region.

Liliaceae

Dichelostemma congestum [*Brodiaea congesta*]
ookow, forktooth ookow, northern saitas, congested snake lily

This native perennial herb generally ranges from 40 to 70 cm tall, arising erect from a scaly, deep-seated corm. Its 2 to 3 leaves are all basal, 3 to 10 mm broad and up to 60 cm long, usually persisting until flowering. Flowers occur on a naked scape and are crowded densely together in a terminal cluster. Each individual flower features 6 tepals, ranging from lavender to bluish-purple, united about their length and spreading slightly to reveal 3 fertile stamens with broad filaments. Ookow gets the name "forktooth" from its additional 3 infertile stamens, which are deeply bifid. This edible geophyte blooms from May to June in grassy meadows, rocky prairies and sagebrush slopes.

Erythronium oregonum ssp. *oregonum*
white fawn lily

The showy blooms of white fawn lily are borne on a nodding stalk up to 30 cm tall and feature 6 tepals with their tips upturned. The 6 exerted stamens surround the compound pistil with a trifid stigma. The leaves are basal and often mottled with light and dark blotches. Fawn lily gets its common name from the leaves' resemblance to the spotted coat of a fawn. This March to April blooming native perennial is found in well-drained open areas and oak woodlands.

Fritillaria affinis [*F. lanceolata*]
chocolate lily

Chocolate lily flowers feature 6 brown and yellow mottled tepals, usually growing 15 to 100 cm tall in nodding clusters of 2 to 5. The parallel-veined leaves are lanceolate, typically whorled near the base of the plant and alternate just below the flowers. This native perennial grows in open areas and in adjoining forest or oak woodland edge habitat. The bulbs of chocolate lily have been and continue to be an important food source for many Native peoples of Puget Sound. It blooms from April to June.

Lilium columbianum
tiger lily

This hairless native perennial grows 60 to 120 cm tall from scaly bulbs. Showy flowers are borne singly or sometimes in pairs on long pedicels. Six reflexed tepals are bright yellow-orange to deep orange with dark red spots near the center and 6 stamens that are prominently exerted. Lanceolate leaves form whorls around the lower portion of the stem. Found in prairies and oak woodlands, it blooms from May to August.

Triteleia grandiflora
[*Brodiaea howellii*]
Howell's brodiaea

Ranging on both sides of the Cascades in dry and well-drained areas such as coastal bluffs, prairies, and sagebrush desert, Howell's brodiaea is a native perennial with white to light or dark blue flowers. Flowers emerge in an open, involucral umbel and have 6 tepals from 18 to 30 mm, becoming connate halfway down their length, with edges that are slightly ruffled and only slightly flared. Stamens are yellow, alternate and 6 in number with 2 sets of filaments of different lengths. Slender and grass-like leaves emerge from corms and number from 1 to 5, generally withering by the time the flowers open. Howell's brodiaea blooms from May to June.

Triteleia hyacinthina [*Brodiaea hyacinthina*]
white brodiaea

This perennial herb grows from a scaly corm with a flowering stalk that ranges in height from 30 to 60 cm. The 1 to 2 basal grass-like leaves have parallel venation and are 3 to 10 mm wide. They range in length from 10 to 40 cm. Flowers are arranged in an umbel and have 6 white tepals, each with a prominent green stripe down the middle. Six fertile anthers occur opposite each tepal and usually are supported by a filament that is triangular and dilated at the base. Seeds are borne in 3-celled capsules. White brodiaea grows in grasslands and blooms from May through July.

Trillium parviflorum
small-flowered trillium

This small-flowered trillium is a showy native perennial herb occuring along prairie edges. The flower sits stalkless above 3 lightly mottled, heart-shaped green leaves born on a single stem. The 3 narrow green sepals subtend 3 white petals on stems 20 to 70 cm high. Far less common than our forest-dwelling western trillium (*Trillium ovatum*), small-flowered trillium is often found under Oregon white oak (*Quercus garryana* var. *garryana*) along with chocolate lily (*Fritillaria affinis*). It can be distinguished from western trillium by its lack of a flower stalk. It blooms in late March to early May.

Zigadenus venenosus
meadow death-camas

This native perennial herb generally grows 20 to 50 cm tall growing from a layered bulb. Long linear leaves form at the base with smaller, bract like leaves forming upward along the stem. The terminal inflorescence is a raceme of white flowers with cream and yellow centers. Meadow death-camas can be found in a variety of ecosystems from coastal prairies, grassy bluffs and hillsides, as well as sagebrush steppes and mountain forests. It ranges from British Columbia south to California and as far east as Colorado and the Dakotas. It blooms from May to July. This taxon was recently renamed *Toxicoscordion venenosum* and reclassified in Melanthiaceae.

Orchidaceae—Orchid Family

Spiranthes romanzoffiana [S. r. var. r.]
hooded ladies'-tresses

A native orchid with white flowers, hooded ladies'-tresses can be found in wet areas on the South Sound prairies. Flowers emerge on dense spikes from 3 to 12 cm long, with 1 to 4 spiraling, vertical rows. Each flower is subtended by a green bract 10 to 20 mm long. This species gets its name from the sepals, which are fused and cover upper petals of the flower like a hood. Leaves are hairless, slender, linear and mostly basal with cauline leaves reduced. A perennial herb growing from 10 to 60 cm, this plant blooms from July to September.

Poaceae [Gramineae]—Grass Family

Agrostis capillaris [A. tenuis]
colonial bentgrass

A perennial grass with a tufted leaves 2 to 5 cm wide, colonial bentgrass grows from 20 to 80 cm in height. The inflorescence is an open, slenderly-branched panicle. The spikelets feature 1 flower each, with the glumes exceeding beyond the lemmas. Peel the leaf away from the culm to see the ligule, a 1 to 3 mm long membrane that marks the juncture between the leaf blade and the sheath. Introduced from Europe as a pasture grass, colonial bentgrass is common in open areas at low elevation. It blooms June to August.

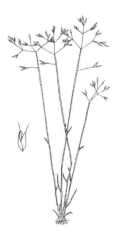

Aira caryophyllea var. *caryophyllea*
[A. c. in part], silver hairgrass

A species introduced from southern Europe, this small annual grass prefers well-drained soils and does well on our local prairies. The very narrow, mostly basal leaves are the feature from which this species gets its common name. The leaves are short, with 1 to 3 mm ligules, lacking auricles. Plants grow from 5 to 25 cm tall and develop compact, spike-like panicles from 1 to 3 cm long with glumes of equal size, slightly longer than the lemmas. The spikelets range in color from silver-green to purplish. Each of the 2 lemmas have a twisted awn, about 2 to 3 mm long.

Aira praecox
little hairgrass

A close relative of silver hairgrass, little hairgrass is similar in size and stature, but has slightly longer, open panicles. Found throughout the western United States, this annual grass has awned lemmas and ligules that are rough to the touch. Both of these grasses bloom in the spring, die, and turn brown as summer progresses.

Anthoxanthum odoratum
sweet vernalgrass

Among our earliest blooming grasses, this tufted perennial gets its common name from its springtime flowering and from the sweet smell it produces. The dense, spike-like inflorescence is borne on a culm reaching 10 to 30 cm and takes on a burgundy color when in flower. Each spikelet includes awned, sterile lemmas and unawned, fertile lemmas. An introduced species of European origin, sweet vernalgrass is widely distributed, growing in meadows, along roadsides and in open forests. It blooms from April to July.

Arrhenatherum elatius
tall oatgrass

This tall, non-native perennial meadow and pasture grass grows most commonly west of the Cascades. The culms of tall oatgrass are hollow, and 80 to 150 cm tall. Leaf blades are 8 to 10 mm wide and often have long hairs. The inflorescence is a panicle, 10 to 30 cm long, with many short branches and florets at each node with bare internodes. Spikelets are 2-flowered, and the paleas are about equal to the lemmas. The awn is twisted and strongly bent where the twisting stops. At the base of the floret is a tuft of hair. It blooms May through July. Both varieties occur in the Puget Sound area, *A. e.* var. *elatius* and *A. e.* var. *bulbosum*.

Bromus carinatus var. *carinatus*
California brome

California brome is a native rhizomatous perennial. Plants can be either hairy or glabrous with hollow culms growing from 30 to 100 cm with closed sheaths and tough leaf blades generally less than 10 mm wide and nearly flat. Ligules are usually from 1 to 3 mm with occasional, tiny auricles. The inflorescence is a narrow, tightly compressed panicle from 10 to 15 cm, branching into several 5 to 10-flowered spikelets 2 to 3 cm long. The glumes and lemmas are keeled; the second glume is nearly twice as long as the first, lemmas are slightly indented at the tip. Awns are straight and 3 to 15 cm long. A readily grazed species, California brome ranges from forest to meadow to sagebrush lands from Alaska to Baja California and east towards the Great Plains. It blooms from May to August.

Bromus sterilis
barren brome, poverty brome

This introduced annual grass is in the fescue tribe. The plant grows 50 to 100 cm in height with flat leaf blades and closed sheaths. The inflorescence is a narrow panicle with a few long, erect branches in each node, holding one inflorescence. Spikelets are 5- to 7-flowered and flattened, 2 to 2.5 cm long, with the prominent awn up to 4 mm. It blooms from May to July.

Danthonia californica
California oatgrass

This native perennial grass in the oat tribe is in bloom from June to July. Plants are 30 to 80 cm high and grow in clumps. Florets are relatively large and sparse with long petioles and lemmas up to 14 mm. The palea is shorter than the lemma, and the lemmas have flattened twisted awns. The lower portion of the leaf blade forms a sheath around the stem, which is hollow. Beyond the sheath, blades are flat and can be hairy or hairless. Its ligule consists of a fairly minor fringe of hair.

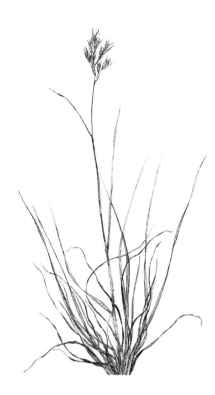

Danthonia intermedia
timber oatgrass

Timber oatgrass is a native perennial bunchgrass distributed throughout the western United States and Canada, especially in the mountains. Highly adaptable, it is shade tolerant and can be found in a wide range of habitats from wet alpine meadows and bogs to dry rocky prairies, alluvial flats, and forests. Erect culms grow in dense tufts from a shallow and fibrous root system, reaching 10 to 50 cm in height. From April to May, they bear narrow, often 1-sided panicles with short, mostly erect branches usually holding 4 to 9 spikelets. Occasionally 1 to 2-flowered spikelets can occur in the axils of old leaves. As the plant ages, the culms often separate at the nodes where these spikelets occur. The leaves are mostly basal with the old blades and sheaths often persistent, withered at the base.

Elymus glaucus ssp. *glaucus*
blue wildrye

This native perennial grass has long stems and compact, flattened spike inflorescences and grows 50 to 100 cm tall. Leaves are borne on the stems with open sheaths, and a few short basal leaves usually dead at maturity. Sheaths have small hairs but leaves and stem are relatively hairless. Florets are long and narrow, packed tightly, 5 to 15 cm long and ornamented by long awns. There are 2 spikelets at each node. It blooms from June to August.

Festuca roemeri
[*F. idahoensis* var. *roemeri*]
Roemer's fescue

Roemer's fescue is a clump-forming perennial bunchgrass that forms a relatively deep root system in contrast with the many turf-forming grasses that spread via rhizomes. Reaching 40 to 100 cm in height, this native grass has an overall blue-gray color. It produces narrow panicles of 5- to 7-flowered spikelets from late May to July. The presence of this fescue indicates dry or very well-drained soils.

Holcus lanatus
common velvet grass

A European introduction, common velvet grass has become a familiar invasive weed. It is easily identified by its grayish color and soft, velvety pubescence on the leaves and culms. A tufted perennial grass growing 50 to 100 cm tall, it has tightly arranged spikelets with 3 flowers. The lower flower is bisexual and the upper flower is male with 3 stamens. The upper lemma has a hooked awn. It blooms from June to September.

Koeleria macrantha
[*K. cristata*]
prairie Junegrass

This native perennial grass blooms from May to July and is found in dry and well-drained locations in sandy and rocky soil in our region. Culms are 30 to 60 cm tall topped with fluffy looking, congested inflorescences (4 to 13 cm) that form a spike-like panicle. Mostly 2-flowered spikelets are borne on short pedicels with paleas shorter than the lemmas. The leaf sheaths are open; leaves are 1 to 2 mm broad and basally tufted.

Lolium multiflorum
Italian ryegrass

This introduced rhizomatous grass is a biennial that grows in tufts 30 to 80 cm tall. Italian ryegrass has hairless stems and leaves, 3 to 4 mm wide. Erect flower spikes have 7-nerved glumes. Its unawned lemma distinguishes it from *L. perenne*, which has an awned lemma. Originally grown for hay, *L. multiflorum* grows in fields, pastures, lawns, and disturbed areas at low elevations. It blooms from May to July.

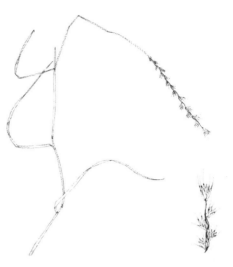

Lolium perenne
perennial ryegrass

A smooth, tufted perennial grass of European origin, perennial ryegrass has hollow culms and grows 30 to 80 cm tall, usually with several short, leafy basal stems. The inflorescence is a single, terminal spike 7 to 25 cm long, with spikelets attached edgewise. It blooms from May to July. Unlike *L. multiflorum*, the lemmas are awned.

Melica subulata var. *subulata*
Alaska oniongrass

Alaska oniongrass is a native tufted perennial with culms that grow up to 1 m tall. It grows from a bulb clustered on short rhizomes. The flat broad leaf blades are slightly rough with scarce hairs on the upper surface. The panicle, a kind of branched inflorescence, is 10 to 20 cm long and narrow. The spikelet is 12 to 24 mm long and contains 2 to 5 flowers. It is found growing in open slopes to thick woods, in dry to moist soils. Its range is from southern Alaska to California and eastward through Idaho, Montana and Wyoming. It blooms from May to July.

Panicum capillare ssp. *capillare*
[*P. c.* in part]
witchgrass, common panicgrass

This native annual grass gets its genus name from the panicle inflorescence that is unusually eye-catching for a grass, reminiscent of fireworks. It prefers moister areas and has a weedy habit. The leaves are pubescent with small white hairs 2 to 4 mm long. Straight hairs form the ligule where the blades are attached to the stem, blades are 5 to 12 mm across—relatively wide. The sheaths are open. The leaves have a visible center vein that is slightly indented. The panicle inflorescence consists of a central stem with branches that increase in length away from the tip. Each branch contains several florets, which sit at the terminus of an unusually long petiole and are tear-drop shaped and rather symmetrical. It grows 20 to 70 cm tall and blooms from June to September.

Panicum occidentale
hairy panicgrass

This tufted native perennial panicgrass grows in open and often disturbed habitats at a variety of elevations throughout Washington. The plant ranges from pale yellow-green to purplish in color. This grass grows to 75 cm tall, with typical plants in our area reaching to 30 cm in height. Leaf blades are broadly lanceolate, 6 to 12 cm in length, with a basal rosette of leaves shorter than those borne on flowering stalks. Leaf blades and culms are variously surfaced with fine hairs. The inflorescence is a broad and open panicle, nearly as wide as it is tall. Spikelets are generally 2-flowered and ± 2 mm long. North American examples of the genus *Panicum* that feature a basal rosette of leaves have been reclassified as members of the genus *Dicanthelium*. Thus the current name for this taxon is *D. acuminatum* ssp. *fasciculatum*.

Phleum pratense
Timothy

A clumping grass that escaped cultivation, Timothy is a short–lived perennial that grows up to 1 m tall, with blades up to 8 mm wide and ligules 2 to 3 mm long. A white to pale green cylindrical panicle, 1 cm by 10 cm, is tightly packed with flowers that have curved awns and hairy glumes. It grows at low to mid-elevations, in pastures, clearings, and near roadsides, blooming from May to July.

Poa annua
annual bluegrass

Reaching only 5 to 20 cm, this diminutive annual grass can form spreading mats by rooting from nodes. The leaf blades are smooth and have prow-like tips; the sheaths closed below, but open for more than half their length. Spikelets are 3 to 6 flowered in spreading to open panicles. Glumes are unequal; the lower 1-nerved glume sits slightly below a broader glume with 5 nerves. This widely distributed grass from Europe grows along roadsides, in open woodlands, in lawns, gardens, and other disturbed areas.

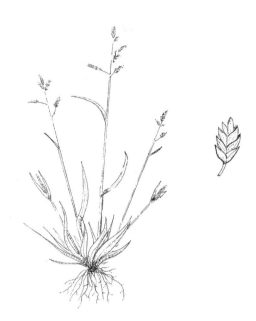

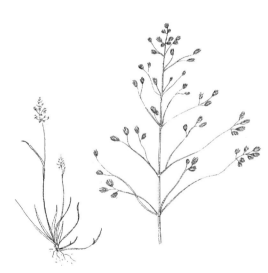

Poa pratensis ssp. *pratensis*
Kentucky bluegrass

Reaching up to 1 m in height, this perennial grass spreads broadly via rhizomes. Loose panicles to 10 cm long host 3 to 5 flowered spikelets. Glumes and lemmas are both strongly keeled and lemmas are cobwebby at the base. Introduced from Europe and common throughout the temperate zones, Kentucky bluegrass grows in open areas from sea-level to mid-elevations. A study of vegetation at Mima Mounds near Littlerock found repeating patterns of particular species of plants occupying particular locations on the mounds (position on or between mounds, aspect, etc.), and Kentucky bluegrass seemed to reach its maximum cover near the tops of the mounds (del Moral and Deardorff 1976).

Appendix A:
List of Voucher Specimens at The Evergreen State College Herbarium

CONIFEROPHYTA. Conifers
Pinaceae. Pine Family
Pinus contorta Dougl. var. *contorta* Shore pine. Glacial Heritage. Constance, Doyle, Hinchliff, & Sheedy 13.
Pseudotsuga menziesii var. *menziesii* (Mirbel) Franco. Douglas-fir. Glacial Heritage, Fort Lewis, Wolf Haven. Brae 4.

ANTHOPHYTA. Flowering Plants

DICTOYLEDONEAE. Plants with Two Seed Leaves
Apiaceae [Umbelliferae]. Carrot Family
Anthriscus caucalis Bieb. *[A. scandicina]* Burr chervil. Glacial Heritage. Constance, Doyle, Hinchliff, & Sheedy 61.
Lomatium nudicaule (Pursh) Coult & Rose. Barestem biscuitroot, pestle parsnip. Fort Lewis. Lombardi *s.n.*
Lomatium triternatum (Pursh) Coult. & Rose var. *triternatum* Nineleaf biscuit root. Glacial Heritage, Fort Lewis. Lombardi *s.n.* and McCain 09-6.
Lomatium utriculatum (Nutt.) Coult. & Rose. Spring gold. Glacial Heritage. Fort Lewis. Scalici 09-4.
Perideridia gairdneri (Hook. & Arn.) Mathias ssp. *borealis* T.I. Chuang & Constance. Common yampah. Glacial Heritage. Elliott 91, 94.

Apocynaceae. Dogbane Family
Apocynum androsaemifolium L. Spreading dogbane. Johnson Prairie, Fort Lewis. Lombardi *s.n.*

Asteraceae [Compositae]. Aster Family
Achillea millefolium L. Yarrow. Glacial Heritage. Elliott 68.
Antennaria howellii Greene ssp. *howellii* *[A. neglecta* var. *howellii* (Greene) Cronq.*]* Field pussytoes, Howell's pussytoes. Tenalquot Prairie. Brae, Treasure, & Politsch 31.
Balsamorhiza deltoidea Nutt. Deltoid balsamroot. Fort Lewis. Lombardi *s.n.*, McCain 09-01.
Cirsium vulgare (Savi) Ten. Bull thistle. Glacial Heritage. Elliott 66.
Crepis capillaris (L.) Wallr. Smooth hawksbeard. Fort Lewis. Lombardi *s.n.*
Erigeron speciosus (Lindl.) DC. Showy fleabane. Fort Lewis. Lombardi *s.n.*
Eriophyllum lanatum (Pursh) Forbes var. *leucophyllum* (DC.) W.R. Carter. Oregon sunshine or common woolly sunflower. Mima Mounds. Klotz *s.n.*
Gnaphalium palustre Nutt. Lowland cudweed. Fort Lewis. Lombardi *s.n.*
Hieracium albiflorum Hook. White-flowered hawkweed. Fort Lewis. Lombardi *s.n.*
Hypochaeris radicata L. Hairy cat's-ear. Fort Lewis. Lombardi. *s.n.*
Logfia minima (Sm.) Dumort. *[Filago minima]* Little cottonrose. Glacial Heritage. Elliott 60.

VASCULAR PLANTS OF THE SOUTH SOUND PRAIRIES 131

Microseris laciniata (Hook.) Schultz-Bip. ssp. *laciniata [M. l.]* Cutleaf microseris. Fort Lewis. Lombardi *s.n.*
Senecio jacobaea L. Tansy ragwort. Glacial Heritage. Elliott 63.
Senecio sylvaticus L. Wood groundsel. Glacial Heritage. Elliott 80.
Sericocarpus rigidus [Aster curtus Cronq.*]* White-topped aster. Fort Lewis, Littlerock. Lombardi *s.n*, Smith *s.n.*
Solidago lepida DC. var. *salebrosa* (Piper) Semple. *[Solidago canadensis* var. *salebrosa]* Rocky Mountains Canada goldenrod. Glacial Heritage. Elliott 77.
Solidago missouriensis var. *tolmieana* Nutt. Missouri goldenrod. Fort Lewis. Lombardi *s.n.*
Solidago simplex Kunth. var. *simplex [S. spathulata* var. *neomexicana]* Mount Albert goldenrod, coast goldenrod Glacial Heritage. Smith 47.
Tanacetum vulgare L. Common tansy. Glacial Heritage. Elliott 65.
Taraxacum erythrospermum Andrz. ex Bess. *[Taraxacum laevigatum* in part*]* Red-seeded dandelion. Fort Lewis. Scalici 09-7.
Taraxacum officinale F.H. Wigg. Common dandelion. Glacial Heritage. Constance, Doyle, Hinchcliff, & Sheedy 4.
Tragopogon dubius Scop. Yellow salsify. Fort Lewis. Lombardi *s.n.*

Berberidaceae. Barberry Family
Vancouveria hexandra (Hook.) Morr. & Dec. White inside-out-flower. Fort Lewis. Lombardi *s.n.*

Boraginaceae. Borage Family
Cynoglossum officinale L. Common hound's-tongue. Pierce County. Giblin 2763.
Myosotis discolor Pers. Yellow and blue forget-me-not. West Rocky Prairie. Brae, Treasure, & Politsch 51.
Myosotis stricta Link ex Roem. & J.A. Schult. *[M. micrantha]* Pall. Blue scorpion-grass. Fort Lewis. McCain 09-9.

Brassicaceae [Cruciferae]. Mustard Family
Arabidopsis thaliana (L.) Schur. Thale cress. Fort Lewis. Pyrooz 09-11.
Arabis eschscholtziana Andrz. *[Arabis hirsute* var. *eschscholtziana]* Pacific coast rockcress. Small prairie in oak forest 1 mile east of Rainier. Barrett 552.
Barbarea orthoceras Ledeb. Yellow rocket. Fort Lewis. Pyrooz 09-15.
Capsella bursa-pastoris (L.) Medic. Shepherd's purse. Fort Lewis. Pyrooz 09-1.
Cardamine hirsuta L. Hairy bittercress. Fort Lewis. Scalici 09-01.
Cardamine oligosperma Nutt. Little western bittercress. Tenalquot Prairie, Wolf Haven. Brae, Treasure, & Politsch 33.
Draba verna L. Spring draba. Fort Lewis. Pyrooz 09-2.
Lepidium campestre (L.) R. Br. Field pepperweed. Fort Lewis. McCain *s.n.*
Teesdalia nudicaulis (L.) R. Br. Barestem teesdalia. Fort Lewis. Lombardi *s.n.* Tenalquot Prairie. Brae, Treasure, & Politsch 19. West Rocky Prairie. Brae, Pickett, Schade 48. Wolf Haven. Brae 5. Fort Lewis. Scalici 09-15.

Campanulaceae. Harebell Family
Campanula rotundifolia L. Common harebell. Fort Lewis. Lombardi *s.n.* Rocky Prairie. Stiles *s.n.*

Caryophyllaceae. Pink Family
Cerastium arvense L. ssp. *strictum* Field chickweed. Fort Lewis. Lombardi

s.n. Tenalquot Prairie. Brae, Treasure, Politsch 28. West Rocky Prairie. Brae, Pickett, Schade 48. Wolf Haven. Brae 3. Fort Lewis. McCain 09-8.
Cerastium glomeratum Thuill. Sticky mouse-ear chickweed. Fort Lewis. Scalici 09-6.
Dianthus deltoides L. Maiden pink. Scatter Creek. Wiedemann s.n.
Sagina decumbens ssp. *occidentalis* (S. Wats.) Crow. Western pearlwort. Fort Lewis. Pyrooz 09-16.
Silene scouleri Hook. ssp. *scouleri*. Scouler's silene. Fort Lewis. Glacial Heritage. Elliott 86. Upper Weir Prairie, Fort Lewis. Lombardi. s.n.
Silene vulgaris (Moench) Garcke [*S. cucubalus*] Bladder campion. Fort Lewis. Lombardi. s.n
Spergularia rubra (L.) Presl. Red sandspurry. Fort Lewis. Pyrooz 09-3.
Stellaria media (L.) Cyrill. Chickweed. Wolf Haven. Brae 8.

Cucurbitaceae. Cucumber Family
Marah oregana (T. & G.) Howell. Coastal manroot. Pierce County. Giblin 2764.

Ericaceae. Heath Family
Arctostaphylos uva-ursi (L.) Spreng. Kinnikinnick. Wolf Haven. Brae 1.

Fabaceae [Leguminosae]. Pea Family
Acmispon nevadensis (S. Wats.) Brouillet var. *n.* [*Lotus n.* var. *douglasii*]. Nevada deervetch. Glacial Heritage. Basey, Devlin, & Schermerhorn 44.
Acmispon parviflorus Benth. [*Lotus micranthus*] Short-flower bird's-foot-trefoil, small-flowered deer vetch. Fort Lewis. Lombardi s.n.
Cytisus scoparius (L.) Link. Scotch broom. Fort Lewis. McCain 09-7.
Lupinus bicolor Lindl. Two-color lupine. Fort Lewis. Lombardi s.n. Pierce County. Giblin 2762.
Lupinus lepidus Dougl. var. *lepidus* Prairie lupine. Fort Lewis. Lombardi s.n.
Trifolium pratense L. Red clover. Mima Prairie. Smith s.n. Glacial Heritage. Elliott 64.
Trifolium subterraneum L. Burrowing clover. Fort Lewis. Pyrooz 09-7.
Vicia americana Muhl. ex Willd. var. *americana* American vetch. Fort Lewis. Lombardi s.n.
Vicia hirsuta (L.) S. F. Gray. Hairy vetch. Fort Lewis. Scalici 09-16.
Vicia sativa L. Common vetch. Fort Lewis. Scalici 09-14.
Vicia sativa L. var. *angustifolia* (L.) Wahlb. Common vetch. Tenalquot Prairie. Brae, Treasure, & Politsch 27. West Rocky Prairie. Brae, Pickett, & Schade 46.
Vicia sativa L. var. *sativa* Common vetch. Fort Lewis. Lombardi s.n.

Gentianaceae. Gentian Family
Centaurium erythraea Rafn. [*Centaurium umbellatum* Gilib. misapplied] European centaury, common centaury. Fort Lewis. Lombardi s.n.

Geraniaceae. Geranium Family
Erodium cicutarium (L.) L'Her. Common stork's-bill. Fort Lewis. Pyrooz 09-6.
Geranium dissectum L. Cut-leaf geranium. Fort Lewis. Pyrooz 09-4.
Geranium molle L. Dovefoot geranium. Fort Lewis. Scalici 09-17.

Hypericaceae (now Clusiaceae). St. John's-wort Family
Hypericum perforatum L. Common St. Johns-wort. Glacial Heritage. Elliott 73.

Mima Prairie. Smith *s.n.*, Gerrish *s.n.*

Lamiaceae [Labiatae]. Mint Family
Lamium purpureum L. Dead-nettle. Fort Lewis. Pyrooz 09-14.
Mentha arvensis L. Corn mint. Fort Lewis. Lombardi *s.n.*
Prunella vulgaris L. var. *lanceolata* (W. Bartram) Fernald. Self-heal. Glacial Heritage. Elliott 52. Fort Lewis. Lombardi *s.n.*

Orobanchaceae. Broomrape Family
Orobanche uniflora L. Naked broomrape. Fort Lewis. Scalici 09-8.

Oxalidaceae. Wood Sorrel Family
Oxalis corniculata L. Creeping yellow wood-sorrel. Fort Lewis. Pyrooz 09-12.

Plantaginaceae. Plantain Family
Plantago lanceolata L. Narrowleaf plantain. Fort Lewis. McCain 09-3.
Plantago major L. Common plantain. Fort Lewis. Lombardi *s.n.*

Plumbaginaceae. Plumbago Family
Armeria maritima (Mill.) Willd. ssp. *californica* (Boiss.) A. E. Porsild. Sea thrift. Tenalquot Prairie. Brae, Treasure, & Politsch 23.

Polemoniaceae. Phlox Family
Microsteris gracilis (Hook.) Greene var. *gracilis*. Slender phlox. Upper Weir Prairie. Lombardi *s.n.*

Polygonaceae. Buckwheat Family
Rumex acetosella L. Sheep sorrel. Tenalquot Prairie. Brae, Treasure, & Politsch 21. Wolf Haven. Brae 15.

Portulacaceae. Purslane Family
Claytonia perfoliata Donn ex Willd. *[Montia perfoliata* in part]. Miner's lettuce. Tenalquot Prairie. Brae, Treasure, & Politsch 32.

Primulaceae. Primrose Family
Dodecatheon hendersonii Gray. Broad-leaved shooting-star. Littlerock. Smith *s.n.* Tenalquot Prairie. Brae, Treasure, & Politsch 25. Wolf Haven. Brae 12.

Ranunculaceae. Buttercup Family
Ranunculus occidentalis Nutt. var. *occidentalis*. Western buttercup. Fort Lewis, Glacial Heritage, Mima Mounds. Scalici 09-9. Ft. Lewis. Pyrooz 09-8.

Rhamnaceae. Buckthorn Family
Ceanothus sanguineus Pursh. Redstem ceanothus. Fort Lewis. Lombardi *s.n.*
Frangula purshiana (DC.) A. Gray *[Rhamnus purshiana]*. Cascara sagrada. Glacial Heritage. Elliott 95.

Rosaceae. Rose Family
Amelanchier alnifolia (Nutt.) Nutt. ex M. Roemer. Serviceberry. West Rocky Prairie. Brae, Pickett, & Schade, 52.
Aphanes australis Rydb. Small-fruited parsley-piert. Fort Lewis. Pyrooz 09-10.
Fragaria vesca L. Woods strawberry. Fort Lewis. Lombardi *s.n.*

Fragaria virginiana Duchesne. Wild strawberry. Fort Lewis. Lombardi *s.n.*
Tenalquot Prairie. Brae, Treasure, Politsch 24. Wolf Haven. Brae 17.
Potentilla gracilis Dougl. var. *gracilis*. Cinquefoil. Rocky Prairie. Stiles *s.n.*
Rubus ursinus Cham. & Schlecht. Trailing blackberry. Littlerock. Smith *s.n.*

Rubiaceae. Madder Family
Galium aparine L. Cleavers. Tenalquot Prairie. Brae, Treasure, & Politsch 26.
Sherardia arvensis L. Blue field madder. Fort Lewis. McCain 09-2.

Saxifragaceae. Saxifrage Family
Heuchera micrantha Dougl. Smallflower alumroot. Fort Lewis. Lombardi *s.n.*
Micranthes integrifolia Hook. Swamp saxifrage. Glacial Heritage. Fort Lewis. Lombardi *s.n.*

Scrophulariaceae. Figwort Family
Castilleja hispida Benth. var. *hispida*. Harsh paintbrush. Fort Lewis. McCain 09-10.
Parentucellia viscosa (L.) Car. Yellow glandweed. Fort Lewis. Lombardi *s.n.*
Synthyris reniformis (Dougl.) Benth. Snow-queen. Fort Lewis. Lombardi *s.n.*, Wolf Haven. Brae 2.
Triphysaria pusilla (Benth.) Chuang & Heckard. [*Orthocarpus pusillus*] Dwarf owl-clover. Wolf Haven. Brae 11.
Veronica arvensis L. Corn speedwell. Fort Lewis. Pyrooz 09-4. McCain 09-4.

Valerianaceae. Valerian Family
Valerianella locusta (L.) Betcke. Lamb's lettuce. Fort Lewis. Scalici 09-11.

Violaceae. Violet Family
Viola adunca Sm. ssp. *adunca* Early blue violet. Glacial Heritage, Mima Mound. Smith *s.n.*, Gerrish *s.n.*
Viola glabella Nutt. Stream violet. Glacial Heritage. Constance, Doyle, Hinchcliff, & Sheedy 33.
Viola howellii A. Gray. Howell's violet. Johnson Prairie. Lombardi *s.n.*
Viola praemorsa Pursh var. *praemorsa*. (Dougl.) Wats. Upland yellow violet. Glacial Heritage. Constance, Doyle, Hinchcliff, & Sheedy 21.

MONOCOTYLEDONEAE. Plants with One Seed Leaf.

Cyperaceae. Sedge Family
Carex inops L. H. Bailey ssp. *inops* [*C. pensylvanica* var. *vespertina*] Long-stoloned sedge. Glacial Heritage. Elliott 51. Mima Prairie. Smith *s.n.*

Iridaceae. Iris Family
Sisyrinchium idahoense E. P. Bicknell. [*S. angustifolium* misapplied in Hitchcock and Cronquist] Blue-eyed grass. Fort Lewis. Lombardi *s.n.*

Juncaceae. Rush Family
Luzula comosa E. Mey. var. *laxa* Buchenau [no Hitchcock and Cronquist synonym] Pacific woodrush. Littlerock. Smith *s.n.* Mima Prairie. Daphne. Tenalquot Prairie. Brae, Treasure, & Politsch 22.
Luzula subsessilis (S. Wats.) Buch. Short-stalked wood-rush Fort Lewis. Scalici. *s.n.*

Liliaceae. **Lily Family**
Brodiaea coronaria (Salisb.) Engl. Harvest brodiaea. Fort Lewis. Lombardi *s.n.*
Camassia quamash (Pursh) Greene ssp. *azurea* (A. Heller) Gould. Common camas. Fort Lewis. Lombardi *s.n.* Tenalquot. Brae, Treasure, & Politsch 18. West Rock Prairie. Brae, Pickett, & Schade 45. Wolf Haven. Brae 14.
Camassia quamash (Pursh) Greene ssp. *maxima* Gould. Dark camas. Fort Lewis. Pyrooz 09-9.
Fritillaria affinis (Schultes) Sealy. Chocolate lily. Fort Lewis. Scalici *s.n.*
Lilium columbianum Leichtlin. Tiger lily. Glacial Heritage. Elliott 55.
Trillium parviflorum V. G. Soukup. Small-flowered trillium. Glacial Heritage. Abair *s.n.*
Zigadenus venenosus Wats. Meadow death-camas. Fort Lewis. Lombardi *s.n.* Current name is *Toxicoscordion venenosum*.

Poaceae [Gramineae]. Grass Family
Agrostis capillaris L. *[A. tenuis]* Colonial bentgrass. Littlerock. Smith *s.n.* Mima Prairie. Smith *s.n.*
Agrostis gigantea Roth. *[A. alba]* Black bentgrass. Glacial Heritage. Elliott 88.
Agrostis pallens Trin. Seashore bentgrass. Glacial Heritage. Elliott 70. Mima Prairie. Gilbert *s.n.*
Alopecurus aequalis Sobol. Shortawn foxtail. Fort Lewis. Lombardi *s.n.*
Anthoxanthum odoratum L. Sweet vernalgrass. Fort Lewis. Lombardi *s.n.* Littlerock. Klotz, Gilbert *s.n.*
Arrhenatherum elatius (L.) Presl. Tall oatgrass. Littlerock. Gilbert *s.n.*
Bromus sterilis L. Barren brome. Fort Lewis. Lombardi *s.n.*
Danthonia californica Bol. California oatgrass. Glacial Heritage. Elliott 53. Johnson Prairie. Lombardi *s.n.* Mima Prairie. Klotz *s.n.* Gilbert *s.n.*
Danthonia spicata (L.) P. Beauv. ex Roem. & Schult. Poverty oatgrass. Upper Weir Prairie. Lombardi *s.n.* Mima Prairie. Smith *s.n.*
Dichanthelium acuminatum (Sw.) Gould & C.A. Clark ssp. *fasciculatum* (Torr.) Freckmann & Lelong. *[Panicum occidentale]* Hairy panicgrass. Mima Mounds. Smith *s.n.*
Elymus glaucus Buckley ssp. *glaucus*. Blue wildrye. Johnson Prairie. Lombardi *s.n.*
Festuca roemeri (Pavlick) Alexeev *[F. idahoensis* var. *roemeri]*. Roemer's fescue. Glacial Heritage. Fort Lewis. Lombardi *s.n.*
Festuca occidentalis Hook. Western fescue. Lombardi *s.n.*
Festuca rubra L. Red fescue. Fort Lewis. Lombardi *s.n.*
Holcus lanatus L. Common velvet grass. Littlerock. Smith *s.n.*
Koeleria macrantha (Ledeb.) Schult (note that *K. cristata* is an illegitimate name). Prairie Junegrass. Mima Prairie. Smith *s.n.*
Lolium multiflorum Lam. Italian ryegrass. Upper Weir Prairie. Lombardi *s.n.*
Lolium perenne L. Perennial ryegrass. Fort Lewis. Lombardi *s.n.*
Panicum capillare L. ssp. *capillare [P. c.* in part*]* Witchgrass. Fort Lewis along Muck Creek. Lombardi *s.n.*
Phleum pratense L. Timothy. Littlerock. Gilbert *s.n.*
Poa annua L. Annual bluegrass. Fort Lewis. Pyrooz *s.n.*
Poa pratensis L. ssp. *pratensis* Kentucky bluegrass. Fort Lewis. Scalici *s.n.*

Note: Family circumscriptions in the book will be updated in the second edition. Some of the classification used from Hitchcock and Cronquist (1973), such as Scrophulariaceae, Orobanchaceae, and Liliaceae, is no longer current.

Appendix B:
Illustrators' Contributions

Yianna Bekris
Camassia quamash (detail)

Joe Bettis
Anemone lyalii
Antennaria howellii ssp. *howellii*
Apocynum androsaemifolium
Carex inops ssp. *inops*
Castilleja hispida var. *hispida*
Cerastium arvense ssp. *strictum*
Chamerion angustifolium
Clinopodium douglasii
Elymus glaucus ssp. *glaucus* (detail)
Frangula purshiana
Hieracium albiflorum
Hieracium scouleri
Iris tenax
Juncus bufonius var. *bufonius*
Lomatium nudicaule
Lupinus lepidus var. *lepidus*
Luzula comosa var. *laxa*
Berberis aquifolium
Orobanche uniflora
Panicum occidentale
Phleum pratense (detail)
Pinus ponderosa var. *ponderosa*
Pseudotsuga menziesii var. *menziesii*
Poa annua
Poa pratensis ssp. *pratensis*
Salix scouleriana
Sericocarpus rigidus
Tragopogon dubius
Trillium parviflorum
Triphysaria pusilla
Viburnum ellipticum

Frederica Bowcutt
Camassia quamash
Lithophragma parviflorum
Triteleia hyacinthina

Emily Driskill
Agrostis capillaris
Balsamorhiza deltoidea
Camassia leichtlinii ssp. *suksdorfii*
Galium aparine
Geranium molle
Koeleria macrantha (detail)
Panicum capillare ssp. *capillare*
Sambucus racemosa var. *racemosa*
Sanicula crassicaulis var. *crassicaulis*
Solidago missouriensis

Natalie Hammerquist
Agoseris grandiflora var. *leptophylla*
Crataegus douglasii
Crataegus monogyna
Dodecatheon hendersonii
Festuca roemeri
Fragaria virginiana
Lamium purpureum
Micranthes integrifolia
Quercus garryana var. *garryana* (leaf detail)

Lisa Hintz
Aira caryophyllea var. *caryophyllea*
Aira praecox
Armeria maritima ssp. *californica*
Brodiaea coronaria
Bromus carinatus var. *carinatus*
Campanula rotundifolia
Castilleja levisecta
Clarkia amoena ssp. *lindleyi*
Claytonia perfoliata
Dodecatheon pulchellum var. *pulchellum*
Elymus glaucus ssp. *glaucus*
Eriophyllum lanatum var. *leucophyllum*
Gaillardia aristata
Galium parisiense
Holcus lanatus

VASCULAR PLANTS OF THE SOUTH SOUND PRAIRIES

Hypochaeris radicata
Koeleria macrantha
Leucanthemum vulgare
Lilium columbianum
Lupinus bicolor
Microseris laciniata ssp. *laciniata*
Parentucellia viscosa
Plagiobothrys scouleri var. *scouleri*
Plantago patagonica
Potentilla gracilis var. *gracilis*
Ranunculus uncinatus
Sherardia arvensis
Solidago simplex var. *simplex*
Spiranthes romanzoffiana
Tellima grandiflora
Triteleia grandiflora
Trifolium dubium
Triodanis perfoliata
Viola glabella
Wyethia angustifolia

Lily Hynson
Anthoxanthum odoratum (detail)
Arrhenatherum elatius (detail)

Krista Koller
Crepis capillaris
Fritillaria affinis
Lolium perenne
Lupinus polyphyllus var. *pallidipes*
Phleum pratense
Vicia americana var. *americana*

Meg Krug
Arctostaphylos uva-ursi
Bromus sterilis
Melica subulata var. *subulata*
Plantago lanceolata

Jordan Marlor
Aquilegia formosa var. *formosa*
Vicia villosa var. *villosa*

Callie Martin
Arrhenatherum elatius
Cytisus scoparius
Danthonia californica
Hypericum perforatum
Plectritis congesta
Rubus ursinus
Taraxacum officinale
Urtica dioica ssp. *gracilis*
Zigadenus venenosus

Irene Matsuoka
Hypericum perforatum (leaf detail)
Pinus contorta var. *contorta* (male cone detail)
Quercus garryana var. *garryana* (acorn detail)

Kathleen McSorley
Acmispon parviflorus
Amelanchier alnifolia
Anaphalis margaritacea
Capsella bursa-pastoris
Cardamine oligosperma
Danthonia intermedia
Delphinium nuttallianum
Draba verna
Lepidium campestre
Lomatium triternatum var. *triternatum*
Lupinus albicaulis
Nuttallanthus texanus
Pinus contorta var. *contorta* (tree)
Plantago major
Prunella vulgaris
Pteridium aquilinum var. *pubescens*
Quercus garryana var. *garryana* (tree form and blossom detail)
Rosa gymnocarpa
Rubus spectabilis
Sisyrinchium idahoense
Teesdalia nudicaulis
Trifolium pratense
Trifolium subterraneum
Veronica arvensis

Viola adunca ssp. *adunca*

Megan Porter
Daucus carota
Erigeron speciosus
Lomatium utriculatum
Marah oregana
Oemleria cerasiformis
Polypodium glycyrrhiza

Stella Rose
Collinsia grandiflora
Collinsia parviflora
Dichelostemma congestum
Marah oregana (fruit)
Rumex acetosella

Brita Zeiler
Achillea millefolium
Anthoxanthum odoratum
Erythronium oregonum ssp. *oregonum*
Lolium multiflorum
Myosotis discolor
Pinus contorta var. *contorta*
Ranunculus occidentalis var. *occidentalis*
Symphoricarpos albus var. *albus*
Viola praemorsa var. *praemorsa*

Glossary

achene: a single-seeded, dry, indehiscent fruit; e.g., a sunflower (*Helianthus annuus*) seed with its shell

aerial: the above ground parts of a plant

aggregate: densely clustered

alternate: node/bud arrangement which alternates position along a stem; buds not arranged opposite each other on a stem

andesite: a fine grained, volcanic rock generally containing abundant plagioclase (feldspar), lesser amounts of hornblende and biotite, and little or no quartz; contains 54 to 62 percent silica; a common lava of volcanic arcs created by subduction; *adj.* andesitic

anther: the expanded, apical, pollen bearing portion of the stamen

apex: the tip or farthest point of a structure from the attachment point

auricle: ear-shaped appendage; e.g., on the base of sheep sorrel leaves (*Rumex acetosella*)

awned: possessing a narrow, bristle-like appendage; e.g., on the florets of many grasses

axial: positioned on or pertaining to the axis

axis: the longitudinal, central supporting structure or line around which various floral parts are borne; e.g., a stem around which leaves emerge

banner: the uppermost of five petals in a papilionaceous flower; e.g., the large upper petal in a pea family (Fabaceae) flower

basal: positioned at or arising from the base; e.g., common dandelion (*Taraxacum officinale*) or common plantain (*Plantago major*) leaves

basalt: a fine grained volcanic rock, typically dark, that contains 45 to 54 percent silica; *adj.* basaltic

biennial: a plant that blooms and fruits in its second year and then typically dies

bi-lobed: divided into two segments; e.g., the petals of spring draba (*Draba verna*)

bisexual: a flower with both male and female reproductive organs

bole: the trunk of a tree

bract: a reduced leaf (or leaf-like structure) at the base of a flower or inflorescence; e.g., the white bracts of a western dogwood (*Cornus nuttallii*) inflorescence

bud: an undeveloped shoot or flower, often encased in scale-like leaves

bulb: an underground shoot derived from fleshy leaves or thickened leaf bases; e.g., underground fleshy leaves of an onion

bur: a structure armed with often hooked or barbed spines or appendages

calyx: all the sepals of a flower, collectively

campanulate: bell-shaped; e.g., the flowers of common harebell (*C. rotundifolia*)

catkin: a dense inflorescence of unisexual flowers lacking petals; e.g., the dangling male inflorescences of red alder

cauline: pertaining to the stem

chlorophyll: a group of green pigments found in plants, algae, and cyanobacteria that aids in photosynthesis

ciliate: a fringe of hairs on the edge of an organ; e.g., on the leaves of Pacific woodrush (*Luzula comosa* var. *laxa*)

compound: with two or more like parts in one organ; e.g., a palmately compound leaf of lupine (*Lupinus* spp.) made up of multiple leaflets

conifer: a cone-bearing plant, producing seeds without a fruit; e.g., Douglas-fir (*Pseudotsuga menziesii* var. *menziesii*)

cordillera: a term for a major mountain range or chain of mountain ranges on a continent

corolla: all the petals of a flower, collectively

culm: a hollow or pithy stem of a grass, sedge, or rush

cyme: a flat or round topped inflorescence, paniculate, in which the terminal flower blooms first.

deciduous: a plant that loses all of its leaves for part of the year

dehiscent: an organ that opens at maturity to release its content, such as a fruit or anther

diamicton: any poorly-sorted deposit, such as a till, a debris flow, or a debris avalanche; also diamict

dioecious: a species with male flowers and female flowers on separate plants

disk/disc florets: a small, tubular flower found in Asteraceae; disk florets make up the center portion of a sunflower (*Helianthus annuus*) inflorescence

dissected: deeply divided into many narrow segments

distal end: the end opposite of the attachment point

drift: a general term for any glacial deposit

elliptical: shaped like a narrow oval

entire margined: not toothed, notched, or divided

epiphytic: a non-parasitic plant that grows on another plant; e.g., licorice fern (*Polypodium glycyrrhiza*) growing on big leaf maple (*Acer macrophyllum*)

equitant: overlapping; e.g., as with the leaves of Oregon iris (*Iris tenax*)

evergreen: a plant that keeps its leaves year-round

feldspar: a common rock forming mineral group consisting of silicates of aluminum, sodium, potassium, and calcium

fiddlehead: the coiled leaf of a young fern

filament: the stalk of the stamen, upholding the anther

floret: an individual flower in an Asteraceae or Poaceae inflorescence; a small flower of a dense cluster

forearc basin: a sedimentary basin, commonly elongated, between a subduction-related trench and a volcanic arc in a convergent plate boundary tectonic setting

gametophyte: gamete-producing generation in the plant reproductive life cycle

geomorphology: the science that treats the general configuration of the Earth's surface, its landforms, and their causes and history; *adj.* geomorphic

glabrous: hairless

gland: a structure that secretes an oily or sticky substance

glomerule: a dense cluster of parts

glumes: the paired bracts below a grass spikelet

haustoria: a root-like organ found in parasitic plants that derives nourishment from host plants

hemiparasitic: a plant that is both photosynthetic and parasitic

herbaceous: non-woody

hirsute: having coarse or stiff hairs

hypanthium: a cup-shaped union of the bottom portions of a flower surrounding the ovary

indehiscent: not opening at maturity

indeterminate: an inflorescence in which the flowers at the base open first and the ones at the tip bloom last

inferior ovary: an ovary that is positioned below the attachment point of the other floral whorls, those whorls appearing to be attached to the top of the ovary

inflorescence: a cluster of florets

internode: distance between two nodes

involucre: a whorl of bracts subtending a flower or flower cluster; 'involucrate' refers to a plant having an involucre

irregular flower: bilaterally symmetrical

keel: a prominent ridge running the length of a structure

lanceolate: much longer than wide

latex: a milky sap

lax: loose; with parts opening and spreading

leaf axil: the upper angle between a leaf/petiole and stem

leaflet: individual segment of a compound leaf

lemma: the lower of two bracts enclosing a grass floret

lidar: an acronym for light detection and ranging, a type of remote sensing that uses infrared radiation scans of a landscape or other surface to precisely map the relief of that surface; also used for the digital products derived from this technology

ligule: the long, flat petal of a ray floret in plants of Asteraceae

lingulate: tongue-shaped

lithosphere: the brittle outer part of the Earth, about 100 km in thickness, that comprises the tectonic plates

lobe: a rounded segment of an organ

megashear: tectonic shearing spread out over a large region

mica: a group of silicate minerals that cleave into thin sheets; *adj.* micaceous, containing mica

monoecious: species with unisexual male and female flowers on the same plant

morphology: aspects of form; e.g., shape, color, structure, pattern

naturalized: a non-native species that now maintains a stable population without human involvement; e.g., oxeye daisy (*Leucanthemum vulgare*)

nectary: nectar-secreting gland on a flower

node: part of the stem where buds and leaves originate

oblanceolate: more narrow than long and attached at the narrowest of the two ends

obovate: egg shaped with the broader end terminal

opposite: pairs on opposite sides of an axis.; e.g., the leaves in Lamiaceae

orbicular: round or circular and flat

outwash: stratified deposits produced by glacial meltwater

ovary: lowest part of the pistil that encases the ovule(s)

ovate: egg-shaped with the narrow end terminal

ovulate: pertaining to or producing immature seeds

ovule: the structure that contains egg(s) and after fertilization becomes the seed

palea: the inner bract of a grass floret often partly covered by the lemma; also refers to chaffy scales on the receptacle of many members of Asteraceae

palmate: leaflets, lobes, or venation of leaves that arise from the same point; e.g., the leaves of a lupine

panicle: a branched, racemose inflorescence with flowers maturing from the bottom upwards

papilionaceous: butterfly-like flowers of Fabaceae with banner, wings, and keel; e.g., pea, lupine, vetch, and Scotch broom flowers

pappus: calyx of flowers of Asteraceae modified into hairs or bristles

parasitic plant: a plant which obtains all or part of its nourishment from another organism; e.g., naked broomrape (*Orobanche uniflora*)

pedicel: stalk of a single flower in an inflorescence

peduncle: stalk of a solitary flower or an entire inflorescence; e.g., the stalk of an Oregon sunshine (*Eriophyllum lanatum* var. *leucophyllum*) flower

perennial: a plant that can survive more than two years

perigynium: a modified bract that encases *Carex* achenes

perigynous disk: an enlargement or outgrowth of the receptacle around the base of the ovary

petiole: leaf stalk

photosynthetic: a plant or parts of a plant which use photosynthesis to produce its energy; photosynthesis is a process in which water, carbon dioxide, and sunlight are chemically altered to form energy for the plant in the form of carbohydrates

pinna: a single division of a pinnately compound leaf; plural pinnae

pinnate: in leaves: a compound leaf consisting of an axis and two rows of leaflets - this is also called pinnately compound; in venation: a central vein with many smaller veins branching off, characteristic of eudicots; e.g., the leaves of vetch (*Vicia* spp.)

pinnatifid: pinnately cleft but the divisions don't reach the central axis or rachis

pistil: the central whorl of a female or bisexual flower capable of producing seeds; includes the stigma, style, and ovary with ovules

plagioclase: a series of feldspar minerals that vary in sodium and calcium content

plate tectonics: the theory that describes the Earth's lithosphere as divided into a number of individual quasi-rigid plates whose horizontal and vertical movements and interactions with other plates give rise to seismicity and volcanism

pome/pomme: a fruit with a soft, fleshy outer layer and harder seed-enclosing center; e.g., an apple (*Malus* sp.)

pseudanthium: an inflorescence consisting of many inconspicuous flowers which combine to resemble a single flower; e.g., the flower of a western dogwood (*Cornus nuttallii*) is an inflorescence with showy bracts resembling petals

pubescent: having hairs, hairy

punctate: spotted with translucent or opaque glands

raceme: an unbranched inflorescence that is indeterminate (the oldest flowers are located towards the base and new flowers are produced as the shoot grows) with individual flowers on short stalks; *adj.* racemose

ray floret: small, tubular flower with a ligulate corolla found in Asteraceae; e.g., ray florets make up the outer circle of petals of the pseudanthium found on sunflower (*Helianthus annuus*)

regular flower: radially symmetrical

reniform: kidney-shaped

rhizome: an underground, horizontal, modified stem that facilitates asexual reproduction; e.g., a bearded German iris (*Iris germanica*), ginger (*Zingiber officinale*), and licorice fern (*Polypodium glycyrrhiza*)

riparian: interface between land and a river, stream, lake or other freshwater hydrological features; e.g., the riparian vegetation dominated by Oregon ash (*Fraxinus latifolia*) found along the Black River

rosette: a circular cluster of typically basal leaves; e.g., leaves of the dandelion (*Taraxacum officinale*)

savanna (also savannah): a grassland habitat with widely spaced trees

scape: a leafless flowering stalk arising from the ground of a stemless plant; e.g., stem of the chocolate lily flower (*Fritillaria affinis*)

scorpioid: a coiled inflorescence with lateral flowers developing alternately on opposite sides of the main axis

sepal: the outermost whorl of a flower, often but not always green in color

serrate: having a toothed margin; e.g., the toothed leaves of serviceberry (*Amelanchier alnifolia*)

sheath: an organ which partly or wholly surrounds another organ; e.g., the sheathing leaf bases of grass

silicle: a dry, bicarpellate fruit of the mustard family (Brassicaceae) that is dehiscent and less than twice as long as wide; e.g., the heart-shaped fruits of shepherd's purse (*Capsella bursa-pastoris*)

silique: a dry, bicarpellate fruit of the mustard family (Brassicaceae) that is dehiscent and usually over twice as long as it is wide; e.g., the long and sleek pods of shotweed (*Cardamine oligosperma*)

sorus: a cluster of spore-producing structures of ferns

spike: a type of raceme in which individual flowers lack pedicels; the term spikelet can refer to a small spike

sporophyte: the spore-producing generation of the plant reproductive cycle; the dominant and conspicuous plant in the vascular species

spur: a hollow appendage of the flower; e.g., the spur on larkspur (*Delphinium* sp.)

stade: substage of a glacial stage that is marked by readvance of the glacier

stamen: the male reproductive organ of a flower, usually consisting of an anther and a filament

stigma: the part of the female structure in a flower capable of receiving pollen located at the top of the pistil

stolon: a creeping stem growing along the surface of the ground which is capable of rooting at its nodes; e.g., the stolons of strawberry (*Fragaria* spp.)

style: the stalk that connects the stigma to the ovary

subduction: the process of one lithospheric plate descending under another

subduction zone: a long narrow belt in which subduction takes place

subtend: to be below and close to

succulent: fleshy, like a cactus

superior ovary: located above the point of attachment of all the other floral whorls, such as sepals, petals, and/or stamens

tannin: an astringent, bitter plant polyphenolic compound that binds to and precipitates proteins and various other organic compounds including amino-acids and alkaloids

taproot: an enlarged, somewhat cylindrical to tapering plant root that grows vertically downward; e.g., a carrot root

tectonic: relating to the individual lithospheric plates, their movements and interactions with other plates; see plate tectonics

tendril: a slender, coiling, or twining organ with which a climbing plant uses to grasp for support; e.g., in wild cucumber (*Marah oregana*)

tepal: the sepals and petals collectively when they are indistinguishable, or nearly so, in form and color; e.g., in chocolate lily (*Fritillaria affinis*)

terminal: occurring at the tip

ternate: occurring in threes

thicket: a dense cluster of shrubby plants

toothed: having small lobes or points

trifoliate: with three leaflets; e.g., the leaves of strawberry (*Fragaria* spp.) or clover (*Trifolium* spp.)

umbel: an inflorescence made up of single flowers radiating from peduncles originating from the same point, characteristic of Apiaceae; an umbellet is the cluster of flowers at one of the ends of a compound umbel

unisexual: a flower with either male or female reproductive organs, but not both

vascular: having conductive tissue or pertaining to plants with conductive tissue

whorled: a ring of three or more similar structures radiating from a common point; e.g., the whorled leaves of a tiger lily (*Lilium columbianum*)

wing: a pair of lateral petals found on papilionaceous flowers

woodland: vegetation dominated by hardwood trees

zygomorphic: bilaterally symmetrical; e.g., a violet (*Viola* spp.) or orchid flower

Note: Many of the definitions are paraphrased or excerpted from Harris and Harris (2001), Hitchcock and Cronquist (1973), and Zomlefer (1994).

Literature Cited

Altman, Bob. 2011. "Historical and Current Distribution and Population of Bird Species in Prairie-Oak Habitats in the Pacific Northwest." *Northwest Science* 85(2):194-222.

Apostol, Dean. 2006. "Ecological Restoration." In *Restoring the Pacific Northwest: The Art and Science of Ecological Restoration in Cascadia*, ed. Dean Apostol and Marcia Sinclair, 11-25. Washington D.C.: Island Press.

Arnett, J. 2015. Washington Rare Plant List, 2015. Washington Natural Heritage Program, Department of Natural Resources, Olympia. January 2015.

Aubrey, Dennis. 2013. "Oviposition Preference in Taylor's Checkerspot Butterflies (*Euphydryas editha taylori*): Collaborative Research and Conservation with Incarcerated Women." Master's thesis, The Evergreen State College, Olympia, Washington.

Barnett, Elizabeth A., Ralph A. Haugerud, Brian L. Sherrod, Craig S. Weaver, Thomas L. Pratt and Richard J. Blakely, compilers. 2010. *Preliminary Atlas of Active Shallow Tectonic Deformation in the Puget Lowland, Washington*. U.S. Geological Survey Open-File Report 2010-1149. pubs.usgs.gov/of/2010/1149/ (accessed Aug. 16, 2010).

Barnosky, Cathy W. 1981. "A Record of Late Quaternary Vegetation from Davis Lake, Southern Puget Lowland, Washington." *Quaternary Research* 16(2):221-239.

_____. 1984. "Late Pleistocene and Early Holocene Environmental History of Southwestern Washington State, U.S.A." *Canadian Journal of Earth Sciences* 21(6):619-629.

_____. 1985. "Late Quaternary Vegetation Near Battle Ground Lake, Southern Puget Trough, Washington." *Geological Society of America Bulletin* 96(2):263-271.

Berg, Andrew W. 1989. "Formation of Mima Mounds: A Seismic Hypothesis." *Geology* 18(3):281-284.

_____. 1990. "Comment and Reply on 'Formation of Mima Mounds: A Seismic Hypothesis,' Reply." *Geology* 18(12):1260-1261.

_____. 1991. "Comment and Reply on 'Formation of Mima Mounds: A Seismic Hypothesis,' Reply." *Geology* 19(3):284-285.

Bevan, Dane. 2004. "Oregon Land Donation Claim Notification." The Oregon Historical Society: The Oregon History Project. www.ohs.org/education/

oregonhistory/historical_records/dspDocument.cfm?doc_ID=2D6350BC-CC5A-E143-2EB7E422E9DFE9A7 (accessed April 30, 2015).

Blumenthal, Richard W. 2009. *Charles Wilkes and the Exploration of Inland Washington Waters: Journals from the Expedition of 1841.* Jefferson, North Carolina: McFarland and Company, Inc.

Booth, Derek B. 1994. "Glaciofluvial Infilling and Scour of the Puget Lowland, Washington, During Ice-sheet Glaciation." *Geology* 22(8):695-698.

Booth, Derek B. and Barry S. Goldstein. 1994. "Patterns and Processes of Landscape Development by the Puget Lobe Ice Sheet." In "Regional Geology of Washington State," conveners, Raymond Lasmanis and E. S. Cheney, 207-218, *Washington Division of Geology and Earth Resources Bulletin* 80. Olympia: Washington State Department of Natural Resources.

Booth, Derek B., Ralph A. Haugerud and Kathy Goetz Troost. 2003. "The Geology of Puget Lowland Rivers." In *Restoration of Puget Sound Rivers*, ed. David R. Montgomery, Susan Bolton, Derek B. Booth and Leslie Wall, 14-45. Seattle: University of Washington Press.

Borden, Richard K. and Kathy Goetz Troost. 2001. "Late Pleistocene Stratigraphy in the South-Central Puget Lowland, Pierce County, Washington." Washington Division of Geology and Earth Resources Report of Investigations 33. Olympia: Washington State Department of Natural Resources. www.dnr.wa.gov/Publications/ger_ri33_late_pleistocene_stratig.pdf (accessed March 6, 2008).

Börstler, Boris, Carsten Renker, Ansgar Kahmen and François Buscot. 2006. "Species Composition of Arbuscular Mycorrhizal Fungi in Two Mountain Meadows with Different Management Types and Levels of Plant Diversity." *Biology and Fertility of Soils* 42:286-298.

Bretz, J. Harlen. 1913. "Glaciation of the Puget Sound Region." *Washington Geological Survey Bulletin* 8:9-244. Olympia: Washington Geological Survey.

Burnham, Jennifer L. Horwath and Donald L. Johnson, eds. 2012. "Mima Mounds: The Case for Polygenesis and Bioturbation." *Geological Society of America Special Paper* 490.

Buschmann, Glen. 1997. "Weeds of the South Puget Sound Prairies." In *Ecology and Conservation of the South Puget Prairie Landscape*, ed. Patrick Dunn and Kern Ewing, 163-178. Seattle: The Nature Conservancy of Washington.

Caldwell, Bruce A. 2006. "Effects of Invasive Scotch Broom on Soil Properties in a Pacific Coastal Prairie Soil." *Applied Soil Ecology* 32(1):149-152.

Camp, Pamela and John G. Gamon, eds. 2011. *Field Guide to the Rare Plants of Washington*. Seattle: University of Washington Press.

Caplow, Florence. 2004. "Reintroduction Plan for Golden Paintbrush (*Castilleja levisecta*)." Report prepared for the United States Fish and Wildlife Service, Western Washington Fish and Wildlife Office. Olympia: Washington Department of Natural Resources Natural Heritage Program. www.dnr.wa.gov/Publications/amp_nh_cale_reintroduction.pdf (accessed May 10, 2015).

Carpenter, Cecilia Svinth. 1986. *Fort Nisqually: A Documented History of Indian and British Interaction*. Tacoma, Washington: Tahoma Research Service.

Center for Natural Lands Management. 2014a. "Mazama pocket gopher (*Thomomys mazama*)." Cascadia Prairie-Oak Partnership. cascadiaprairieoak.org/working-groups/mazama-pocket-gopher (accessed May 1, 2015).

_____. 2104b. "Streaked Horned Lark (*Eremophila alpestris strigata*)." Cascadia Prairie-Oak Partnership. cascadiaprairieoak.org/working-groups/streaked-horned-lark (accessed May 1, 2015).

_____. 2014c. "Taylor's checkerspot butterfly (*Euphydryas editha taylori*) 2013 Action Plan Summary." Cascadia Prairie-Oak Partnership. cascadiaprairieoak.org/working-groups/taylors-checkerspot-butterfly (accessed May 1, 2015).

Chappell, Christopher B. 2006. "Upland Plant Associations of the Puget Trough Ecoregion, Washington." Natural Heritage Report 2006-01. Olympia: Washington Department of Natural Resources, Natural Heritage Program.

Chappell, Christopher B. and Rex C. Crawford. 1997. "Native Vegetation of the South Puget Sound Prairie Landscape." In *Ecology and Conservation of the South Puget Prairie Landscape*, ed. Patrick Dunn and Kern Ewing, 107-122. Seattle: The Nature Conservancy of Washington.

Clampitt, Christopher A. 1993. "Effects of Human Disturbances on Prairies and the Regional Endemic *Aster curtus* in Western Washington." *Northwest Science* 67(3):163-169.

Consortium of Pacific Northwest Herbaria. 2007-2013. www.pnwherbaria.org (accessed May 10, 2015).

Constance, Anna, Vinson Doyle, Cody Hinchliff and Rebecca Sheedy. 2003. "A Preliminary Study of Glacial Heritage Preserve, Thurston County, Washington." Unpublished manuscript. The Evergreen State College,

Olympia, Washington.

Cooper, James Graham. 1994. *Plant Life of Washington Territory: Northern Pacific Railroad Survey, Botanical Report, 1853-1861.* Reprinted with added papers by Nelsa M. Buckingham, ed. Alice Racer Anderson. *Douglasia Occasional Papers* Vol. 5. Woodinville: Washington Native Plant Society.

Cramer, M. and N. Barger. 2014. "Are Mima-like Mounds the Consequence of Long-term Stability of Vegetation Spatial Patterning?" *Palaeogeography, Palaeoclimatology, Palaeoecology* 409:72-83.

Crandell, Dwight R. 1963. "Surficial Geology and Geomorphology of the Lake Tapps Quadrangle, Washington." *U.S. Geological Survey Professional Paper* 388-A. pubs.usgs.gov/pp/0388a/report.pdf (accessed August 15, 2015).

Crawford Rex C. and Heidi Hall. 1997. "Changes in the South Puget Sound Prairie Landscape." In *Ecology and Conservation of the South Puget Prairie Landscape*, ed. Patrick Dunn and Kern Ewing, 11-15. Seattle: The Nature Conservancy of Washington.

Cwynar, Les C. 1987. "Fire and Forest History of the North Cascade Range." *Ecology* 68(4):791-802.

Davis, M. B. 1973. "Pollen Evidence of Changing Land Use Around the Shores of Lake Washington." *Northwest Science* 47(3):133-148.

Del Moral, Roger and David C. Deardorff. 1976. "Vegetation of the Mima Mounds, Washington State." *Ecology* 57(3):520-530.

Dennehy, Casey, Edward R. Alverson, Hannah E. Anderson, David R. Clements, Rod Gilbert and Thomas N. Kaye. 2011. "Management Strategies for Invasive Plants in Pacific Northwest Prairies, Savannas and Oak Woodlands." *Northwest Science* 85(2):329-351.

Devine, Warren D. and Constance A. Harrington. 2006. "Changes in Oregon White Oak (*Quercus garryana* Dougl. ex Hook.) Following Release from Overtopping Conifers." *Trees* 20(6):747-756.

Dunn, Patrick. 1998. "Prairie Habitat Restoration and Maintenance on Fort Lewis and within the South Puget Sound Prairie Landscape." Final Report and Summary of Findings for the United States Army, Fort Lewis, Washington. Seattle: The Nature Conservancy of Washington.

Dunwiddie, Peter W. 1987. "Macrofossil and Pollen Representation of Coniferous Trees in Modern Sediments from Washington." *Ecology* 68(1):1-11.

──────. 2009. "Evaluating Suitability of Prairies for Golden Paintbrush (*Castilleja levisecta*) Recovery by Experimental Outplanting in South Puget Sound, Final Report." Seattle, Washington: The Nature Conservancy.

Dunwiddie, Peter, Ed Alverson, Amanda Stanley, Rod Gilbert, Scott Pearson, Dave Hays, Joe Arnett, Eric Delvin, Dan Grosboll and Caroline Marschner. 2006. "The Vascular Plant Flora of the South Puget Sound Prairies, Washington, USA." *Davidsonia* 17(2):51-69.

Dunwiddie, Peter W., Edward R. Alverson, R. Adam Martin and Rod Gilbert. 2014. "Annual Species in Native Prairies of South Puget Sound, Washington." *Northwest Science* 88(2):94-105.

Easterbrook, Don J. 1994a. "Chronology of Pre-Late Wisconsin Pleistocene Sediments in the Puget Lowland, Washington." In "Regional Geology of Washington State," conveners Raymond Lasmanis and E. S. Cheney, 191-206, *Washington Division of Geology and Earth Resources Bulletin* 80. Olympia: Washington State Department of Natural Resources. www.dnr.wa.gov/Publications/ger_b80_regional_geol_wa_2.pdf (accessed April 24, 2013).

_____. 1994b. "Stratigraphy and Chronology of Early to Late Pleistocene Glacial and Interglacial Sediments in the Puget Lowland, Washington." In "Geologic Field Trips in the Pacific Northwest," ed. D. A. Swanson and R. A. Haugerud, 1J 1 - 1J 38. *University of Washington Department of Geological Sciences* Vol. 1.

Easterly, R., D. Salstrom and C. Chappell. 2005. "Wet Prairie Swales of the South Puget Sound." Salstrom and Easterly Eco-logic (SEE) Botanical Consulting in partnership with the Washington Department of Natural Resources Natural Heritage Program.

Ficken, Robert E. 2002. *Washington Territory*. Pullman: Washington State University Press.

Flematti, Gavin R., Emilio L. Ghisalberti, Kingsley W. Dixon and Robert D. Trengove. 2004. "A Compound from Smoke that Promotes Seed Germination." *Science* 305(5686):977.

Foster, Jeffrey R. and Scott E. Shaff. 2003. "Forest Colonization of Puget Lowland Grasslands of Fort Lewis, Washington." *Northwest Science* 77(4):283-296.

Franklin, Jerry F. and C. T. Dyrness. 1973. *Natural Vegetation of Oregon and Washington*. U.S. Forest Service General Technical Report PNW-8. Portland, Oregon: Pacific Northwest Forest and Range Experiment Station.

Freed, S. 2014. "Glacial Heritage Preserve 2014 Pesticide Application Report." Produced for the Washington Department of Health. Olympia, Washington: Center for Natural Lands Management.

Giblin, David. 1997. "*Aster curtus*: Current Knowledge of its Biology and Threats to its Survival." In *Ecology and Conservation of the South Puget Sound Prairie Landscape*, ed. Patrick Dunn and Kern Ewing, 93-100. Seattle, Washington: The Nature Conservancy.

Giovannini, G. and S. Lucchesi. 1997. "Modifications Induced in Soil Physicochemical Parameters by Experimental Fires at Different Intensities." *Soil Science* 162(7):479-486.

Goldstein, B. S., Patrick Pringle, B. Parker and Z. O. Futornick. 2010. "Tracking the Late-Glacial Outburst Flood from Glacial Lake Carbon, Washington State, USA." Northwest Scientific Association, Annual Meeting, 82nd, p. 65.

Gomberg, Joan, Brian Sherrod, Craig Weaver and Art Frankel. 2010. "A Magnitude 7.1 Earthquake in the Tacoma Fault Zone–A Plausible Scenario for the Southern Puget Sound Region, Washington." U.S. Geological Survey Fact Sheet 2010-3023. pubs.usgs.gov/fs/2010/3023/ (accessed May 20, 2010).

Goodrich, Anita. 2013. "Developing Protocols for Establishing a Restoration Nursery." Annual report submitted to the Department of Defense Legacy Resource Management Program. Olympia, Washington: The Center for Natural Lands Management. Project #06 and 07-326. www.dodlegacy.org/Legacy/project/productdocs/06%20and%2007-326%20FS_To%20establish%20a%20series%20of%20permanent%20seed-source%20nursery_0e8c2ca5-e033-4cd0-b8a7-430ac644e880.pdf (accessed May 4, 2015).

Gray, Mary A. 1930. "Settlement of the Claims in Washington of the Hudson's Bay Company and the Puget's Sound Agricultural Company." *The Washington Historical Quarterly* 21(2):95-102.

Grove, Sarah, Karen A. Haubensak and Ingrid Parker. 2012. "Direct and Indirect Effects of Allelopathy in the Soil Legacy of an Exotic Plant Invasion." *Plant Ecology* 213:1869-1882.

Gunther, Erna. 1945. *Ethnobotany of Western Washington: The Knowledge and Use of Indigenous Plants by Native Americans.* Seattle: University of Washington Press.

Hall, Heidi L., Rex Crawford and B. Stephens. 1995. "Regional Inventory of Prairies in the Southern Puget Trough: Phase I." File Report: Washington Department of Natural Resources, Natural Heritage Program, Olympia, Washington: 1-9.

Hamman, Sarah, Jonathan D. Bakker and Sierra Smith. 2015. Regional Prairie

Native Seed Project. Final Report prepared for the U.S. Fish and Wildlife Service.

Hamman, Sarah T., Peter W. Dunwiddie, Jason L. Nuckols and Mason McKinley. 2011. "Fire as a Restoration Tool in Pacific Northwest Prairies and Oak Woodlands: Challenges, Successes, and Future Directions." *Northwest Science* 85(2):317-328.

Hansen, Henry P. 1938. "Postglacial Forest Succession and Climate in the Puget Sound Region." *Ecology* 19:528-548.

_____. 1947. "Postglacial Forest Succession, Climate, and Chronology in the Pacific Northwest." *Transactions of the American Philosophical Society* 37(1):1-130.

Harris, James G. and Melinda Woolf Harris. 2001. *Plant Identification Terminology: An Illustrated Glossary, Second Edition.* Spring Lake, Utah: Spring Lake Publishing.

Haubensak, Karen A. and Ingrid M. Parker. 2004. "Soil Changes Accompanying Invasion of the Exotic Shrub *Cytisus scoparius* in Glacial Outwash Prairies of Western Washington (USA)." *Plant Ecology* 175(1):71-79.

Henry, Erica H. and Cheryl B. Shultz. 2012. "A First Step Towards Successful Conservation: Understanding Local Oviposition Site Selection of an Imperiled Butterfly, Mardon Skipper." *Journal of Insect Conservation* 17:183-194.

Higgs, Eric. 2003. *Nature by Design: People, Natural Process, and Ecological Restoration.* Cambridge, Massachusetts: MIT.

Hill, Kathryn C. and Dylan G. Fischer. 2014. "Native–Exotic Species Richness Relationships Across Spatial Scales in a Prairie Restoration Matrix." *Restoration Ecology* 22(2):204-213.

Hitchcock, C. Leo and Arthur Cronquist. 1973. *Flora of the Pacific Northwest: An Illustrated Manual.* Seattle: University of Washington Press.

Hosten, Paul E., O. Eugene Hickman, Frank K. Lake, Frank A. Lang and David Vesley. 2006. "Oak Woodlands and Savannas." In *Restoring the Pacific Northwest: The Art and Science of Ecological Restoration in Cascadia*, ed. Dean Apostol and Marcia Sinclair, 63-96. Washington D.C.: Island Press.

Hulting, Andrew Gerald, Karin Neff, Eric M. Coombs, Robert Parker, Glenn Miller and L. C. Burrill. 2008. "Scotch Broom: Biology and Management in the Pacific Northwest." Corvallis: Oregon State University, Extension Service; Pullman: Washington State University Cooperative Extension; Moscow: University of Idaho Cooperative Extension System; Washington,

D.C.: U.S. Dept. of Agriculture.

Integrated Taxonomic Information System (ITIS). 2015. www.itis.gov (accessed May 1, 2015).

Kaplan, Mary Jane. 2003. "Introduced Species Summary Project: Scotch Broom (*Cytisus scoparius*)." New York: Columbia University. www.columbia.edu/itc/cerc/danoff-burg/invasion_bio/inv_spp_summ/Cytisus_scoparius.html (accessed May 20, 2014).

Kaye, Thomas N. and Matt Blakeley-Smith. 2008. "An Evaluation of the Potential for Hybridization Between *Castilleja levisecta* and *C. hispida*." Washington Department of Natural Resources, Olympia and Institute for Applied Ecology, Corvallis, Oregon.

Kirkpatrick, H. Elizabeth and Kaitlin C. Lubetkin. 2011. "Responses of Native and Introduced Plant Species to Sucrose Addition in Puget Lowland Prairies." *Northwest Science* 85(2):255-268.

Kruckeberg, Arthur R. 1991. *Natural History of Puget Sound Country*. Seattle: University of Washington Press.

———. 2002. *Geology and Plant Life: the Effects of Landforms and Rock Types on Plants*. Seattle, Washington: University of Washington Press.

Küchler, A. W. 1946. "The Broadleaf Deciduous Forests of the Pacific Northwest." *Annals of the Association of American Geographers* 36(2):122-147.

Lang, F. A. 1961. "A Study of Vegetation Change on the Gravelly Prairies of Pierce and Thurston Counties, Western Washington." Diss. University of Washington, Seattle, Washington.

Lawrence, Beth A. and Thomas N. Kaye. 2006. "Habitat Variation Throughout the Historic Range of Golden Paintbrush, a Pacific Northwest Prairie Endemic: Implications for Reintroduction." *Northwest Science* 80(2):140-152.

Leopold, Estella B., Rudy Nickmann, John I. Hedges and John R. Ertel. 1982. "Pollen and Lignin Records of Late Quaternary Vegetation, Lake Washington." *Science* 218(4579):1305-1307.

Leopold, Estella B. and Robert Boyd. 1999. "An Ecological History of Old Prairie Areas in Southwestern Washington." In *Indians, Fire and the Land in the Pacific Northwest*, ed. Robert Boyd, 139-163. Corvallis: Oregon State University Press.

Logan, Robert L., Timothy J. Walsh, Benjamin W. Stanton and Isabelle Y. Sarikhan. 2009. "Geologic Map of the Maytown 7.5-minute Quadrangle, Thurston County, Washington." Washington Division of Geology and

Earth Resources Geologic Map GM-72. Olympia: Washington State Department of Natural Resources. www.dnr.wa.gov/Publications/ger_gm72_geol_map_maytown_24k.pdf (accessed Apr. 3, 2009).

MacDougall, Andrew. 2002. "Invasive Perennial Grasses in *Quercus garryana* Meadows of Southwestern British Columbia: Prospects for Restoration." In *Proceedings of the Fifth Symposium on Oak Woodlands: Oaks in California's Challenging Landscape*, ed. Richard B. Standiford, Douglas McCreary and Kathryn Purcell, 159-168. U.S. Forest Service General Technical Report PSW-GTR-184. Albany, California: Pacific Southwest Research Station.

Merchant, Carolyn. 2007. *American Environmental History: An Introduction.* New York: Columbia University Press.

Messmer, Louis W. 1994. "Introduction." In *Plant Life of Washington Territory: Northern Pacific Railroad Survey, Botanical Report, 1853-1861*, ed. Alice Racer Anderson. Douglasia Occasional Papers Vol. 5, vi-vii. Woodinville: Washington Native Plant Society.

Mitchell, Rachel M. and Jonathan D. Bakker. 2011. "Carbon Addition as a Technique for Controlling Exotic Species in Pacific Northwest Prairies." *Northwest Science* 85(2):247-254.

Moen, Wayne S. 1986. "Mineral Resource Maps of Washington." Washington State Department of Natural Resources, GM-22:1-4 + maps 1-4. www.dnr.wa.gov/publications/ger_gm22_min_res_wastate.pdf (accessed May 1 2015).

Natural Resources Conservation Service (NRCS). n.d. "Published Soil Surveys for Washington." United States Department of Agriculture. www.nrcs.usda.gov/wps/portal/nrcs/surveylist/soils/survey/state/?stateId=WA (accessed May 25, 2015).

NatureServe. 2015. "An Online Encyclopedia of Life." explorer.natureserve.org (accessed April 21, 2015).

Norton, Helen. 1979. "The Association Between Anthropogenic Prairies and Important Food Plants in Western Washington." *Northwest Anthropological Research Notes* 13(2):175-200.

Pearson, Scott F. and Bob Altman. 2005. "Range-wide Streaked Horned Lark (*Eremophila alpestris strigata*) Assessment and Preliminary Conservation Strategy." Washington Department of Fish and Wildlife, Olympia.

Perdue, Van. 1997. "Land-Use and the Ft. Lewis Prairies." In *Ecology and Conservation of the South Puget Prairie Landscape*, ed. Patrick Dunn and Kern Ewing, 17-28. Seattle: The Nature Conservancy of Washington.

Perry, Laura G., Dana M. Blumenthal, Thomas A. Monaco, Mark W. Paschke and Edward F. Redente. 2010. "Immobilizing Nitrogen to Control Plant Invasion." *Oecologia* 163:13-24.

Porter, Sasha. 2014. "Mycorrhizal and Microbial Inoculation Affect the Growth of Native Plants Raised for Restoration." Master's thesis, The Evergreen State College, Olympia, Washington.

Potter, Lee Ann and Wynell Schamel. 1997. "The Homestead Act of 1862." *Social Education* 61(6):359-364.

Pringle, Patrick T. and Barry S. Goldstein. 2002. "Deposits, Erosional Features, and Flow Characteristics of the Late-Glacial Tanwax Creek-Ohop Creek Valley Flood—A Likely Source for Sediments Composing the Mima Mounds, Puget Lowland, Washington." Geological Society of America Abstracts with Programs 34(5):A-89. gsa.confex.com/gsa/2002CD/finalprogram/abstract_35127.htm (accessed on June 7, 2008).

Pringle, Russell F. 1990. *Soil Survey of Thurston County, Washington*. U.S. Soil Conservation Service. soildatamart.nrcs.usda.gov/Manuscripts/WA067/0/wa067_text.pdf (accessed on August 29, 2013).

Puget Sound Institute. 2012-2015. "Puget Sound's Climate." Encyclopedia of Puget Sound. University of Washington Tacoma Center for Urban Waters. www.eopugetsound.org/articles/puget-sounds-climate (accessed May 10, 2015).

Raven, Peter H., Ray F. Evert and Susan E. Eichhorn. 2005. *Biology of Plants, Seventh Edition*. New York: W. H. Freeman and Company.

Riddle, Margaret. 2010. "Donation Land Claim Act, Spur to American Settlement of Oregon Territory, Takes Effect on September 27, 1850." Seattle: History Link: The Free Online Encyclopedia of Washington State History. www.historylink.org/index.cfm?DisplayPage=output.cfm&file_id=9501 (accessed May 1, 2015).

Rook, Erik J., Dylan G. Fischer, Rebecca D. Seyferth, Justin L. Kirsch, Carri J. LeRoy and Sarah Hamman. 2011. "Responses of Prairie Vegetation to Fire, Herbicide, and Invasive Species Legacy." *Northwest Science* 85(2):288-302.

Schmidt, Inger. 1997. "Fort Lewis Integrated Training Area Management Program." In *Ecology and Conservation of the South Puget Prairie Landscape*, ed. Patrick Dunn and Kern Ewing, 216-269. Seattle: The Nature Conservancy of Washington.

Schwantes, Carlos A. 1989. *The Pacific Northwest: An Interpretive History*. Lincoln: University of Nebraska Press.

Scott, James and Roland L. DeLorme. 1988. *Historical Atlas of Washington.* Norman: University of Oklahoma Press.

Senos, René, Frank K. Lake, Nancy Turner and Dennis Martinez. 2006. "Traditional Ecological Knowledge and Restoration Practice." In *Restoring the Pacific Northwest*, ed. Dean Apostol and Marcia Sinclair, 393-426. Washington, D.C.: Island Press.

Sherrod, Brian L. 2001. "Evidence for Earthquake-Induced Subsidence about 1100 yr Ago in Coastal Marshes of Southern Puget Sound, Washington." *Geological Society of America Bulletin* 113(10):1299-1311.

Sinclair, Marcia, Ed Alverson, Patrick Dunn, Peter Dunwiddie and Elizabeth Gray. 2006. "Bunchgrass Prairies." In *Restoring the Pacific Northwest: The Art and Science of Ecological Restoration in Cascadia*, ed. Dean Apostol and Marcia Sinclair, 29-62. Washington D.C.: Island Press.

Snowden, Clinton A. 1909. *The History of Washington: The Rise and Progress of an American State, Volume 2.* New York: The Century History Company.

Sprenger, Samantha Martin. 2006. "Golden Paintbrush (*Castilleja levisecta*)." Plant data sheet produced for University of Washington's Native Plant Production course website. depts.washington.edu/propplnt/Plants/Clevisecta.htm (accessed May 10, 2015).

Stanley, Amanda G., Thomas N. Kaye and Peter W. Dunwiddie. 2008. "Regional Strategies for Restoring Invaded Prairies: Observations from a Multisite, Collaborative Research Project." *Native Plants Journal* 9(3):247-254.

Stinson, Derek W. 2005. "Washington State Status Report for the Mazama Pocket Gopher, Streaked Horned Lark, and Taylor's Checkerspot." Washington Department of Fish and Wildlife, Olympia. wdfw.wa.gov/publications/00390/ (accessed May 10, 2015).

Storm, L. 2004. "Prairie Fires and Earth Mounds: The Ethnoecology of Upper Chehalis Prairies." *Douglasia* 28(3):6-9.

Storm, Linda and Daniela Shebitz. 2006. "Evaluating the Purpose, Extent, and Ecological Restoration Applications of Indigenous Burning Practices in Southwestern Washington." *Ecological Restoration* 24(4):256-268.

Suding, Katharine N., Katherine D. LeJeune and Timothy R. Seastedt. 2004. "Competitive Impacts and Responses of an Invasive Weed: Dependencies on Nitrogen and Phosphorus Availability." *Oecologia* 141(3):526-535.

Sudworth, George B. 1967. *Forest Trees of the Pacific Slope.* New York: Dover Publications.

Thomas, Ted B. and Andrew B. Carey. 1996. "Endangered, Threatened, and

Sensitive Plants of Fort Lewis, Washington: Distribution, Mapping, and Management Recommendations for Species Conservation." *Northwest Science* 70(2):148-163.

Thorpe, Andrea S. and Amanda G. Stanley. 2011. "Determining Appropriate Goals for Restoration of Imperiled Communities and Species." *Journal of Applied Ecology* 48(2):275-279.

Troost, Kathy Goetz. 2007. "Jökulhlaups from Glacial Lake Puyallup, Pierce County, Washington." *Geological Society of America Abstracts with Programs* 39(4)13. gsa.confex.com/gsa/2007CD/finalprogram/abstract_121416.htm (accessed on August 31, 2013).

Troost, Kathy Goetz and Derek B. Booth. 2008. "Geology of Seattle and the Seattle Area, Washington." In "Landslides and Engineering Geology of the Seattle, Washington Area," ed. Rex L. Baum, Jonathan W. Godt and Lynn M. Highland, 1-35, *Geological Society of America Reviews in Engineering Geology* XX. www.wou.edu/las/physci/taylor/g473/seismic_hazards/troost_booth_2008_geo_seattle.pdf (accessed April 23, 2013).

Troxel, Kathryn Marie. 1950. *Fort Nisqually and the Puget's Sound Agricultural Company.* Bloomington: Indiana University.

Tsukada, Matsuo and Shinya Sugita. 1982. "Late Quaternary Dynamics of Pollen Influx at Mineral Lake, Washington." *Botanical Magazine* 95:401-418.

Tsukada, Matsuo, Shinya Sugita and D. M. Hibbert. 1981. "Paleoecology of the Pacific Northwest. I. Late Quaternary Vegetation and Climate." *Verhandlungen der Internationale Vereinigung für Theoretische und Angewandte Limnologie* 21(2):730-737.

Tveten, Richard. 1997. "Fire Effects on Prairie Vegetation Fort Lewis, Washington." In *Ecology and Conservation of the South Puget Prairie Landscape*, ed. Patrick Dunn and Kern Ewing, 123-130. Seattle: The Nature Conservancy of Washington.

United States Census Bureau. n.d. www.census.gov/population/www/censusdata/pop1790-1990.html (accessed May 1, 2015).

United States Fish and Wildlife Service. 2012. "Endangered and Threatened Wildlife and Plants: Listing Taylor's Checkerspot Butterfly and Streaked Horned Lark and Designation of Critical Habitat; Proposed Rule." 77(197):61938-62058. www.fws.gov/policy/library/2012/2012-24465.html (accessed May 10, 2015).

United States Fish & Wildlife Service (USFWS). 2014. "Designation of Critical Habitat for Mazama Pocket Gophers; Final Rule." *Federal Register* 79 FR

19711 19757.

Walsh, T. J., M. A. Korosec, W. M. Phillips, R. L. Logan, H. W. Schasse, K. L. Meagher and R. A. Haugerud. 1999. Geologic Map of Washington—Southwest Quadrant (digital edition). U.S. Geological Survey Open-file Report 99-382, version 1.0.

Walsh, Timothy J., Michael Polenz, Robert L. Logan, Marvin A. Lanphere and Thomas W. Sisson. 2003. "Pleistocene Tephrostratigraphy and Paleogeography of Southern Puget Sound near Olympia, Washington." In "Western Cordillera and Adjacent Areas," ed. Terry W. Swanson, 225-236, *Geological Society of America Field Guide* Vol. 4. fieldguides.gsapubs.org/content/by/year (accessed August 15, 2015).

Walsh, Timothy J. and Robert L. Logan. 2005. "Geologic Map of the East Olympia 7.5-minute Quadrangle, Thurston County, Washington." Washington Division of Geology and Earth Resources Geologic Map GM-56, scale 1:24,000. Olympia: Washington State Department of Natural Resources. www.dnr.wa.gov/Publications/ger_gm56_geol_map_eastolympia_24k.pdf (accessed March 6, 2008).

Washburn, A. L. 1988. "Mima Mounds: An Evaluation of Proposed Origins with Special Reference to the Puget Lowlands." Washington Division of Geology and Earth Resources Report of Investigations 29. Olympia: Washington State Department of Natural Resources. www.dnr.wa.gov/publications/ger_ri29_mima_mounds.pdf (accessed Feb. 18, 2010).

Washington Department of Fish and Wildlife. 2013. "Threatened and Endangered Wildlife in Washington: 2012 Annual Report." Listing and Recovery Section, Wildlife Program, Washington Department of Fish and Wildlife, Olympia. wdfw.wa.gov/publications/01542/ (accessed May 10, 2015).

_____. 2014. "Mazama Pocket Gopher: Frequently Asked Questions." wdfw.wa.gov/publications/01215/ (accessed February 20, 2014).

Washington Office of the Secretary of State, n.d. www.sos.wa.gov/legacy/images/maps/jpg/AR_mapPSAgricCo1855a.jpg (accessed May 1, 2015).

Weiser, Andrea and Dana Lepofsky. 2009. "Ancient Land Use and Management of Ebey's Prairie, Whidbey Island, Washington." *Journal of Ethnobiology* 29(2):184-212.

Western Regional Climate Center. 2015. "Climate of Washington." Reno, Nevada. www.wrcc.dri.edu/narratives/washington/ (accessed May 10, 2015).

_____. n.d. "Washington Climate Summaries." Reno, Nevada. www.wrcc.dri.

edu/summary/climsmwa.html (accessed May 1, 2015).

White, Richard. 1980. *Land Use, Environment, and Social Change: The Shaping of Island County, Washington.* Seattle: University of Washington Press.

Whitlock, Cathy. 1992. "Vegetational and Climatic History of the Pacific Northwest during the Last 20,000 Years: Implications for Understanding Present-day Biodiversity." *Northwest Environmental Journal* 8:5-28.

Zomlefer, Wendy B. 1994. *Guide to Flowering Plant Families.* Chapel Hill, North Carolina: The University of North Carolina Press.

Index

Page numbers in *italic* refer to illustrations or tables. Synonyms are in brackets.

7S Prairie, 5
13th Division Prairie, 5, *6*
91st Division Prairie, 5

A

Abies (fir), 8-9
Achillea millefolium (yarrow), 25, 33, 60, 70, 130, 138
Acmispon nevadensis var. *nevadensis* (Nevada deervetch), 132
Acmispon parviflorus [Lotus micranthus], 61, 83, 132, 137
acorn consumption by Native Americans, 89
African-American settlers, 47
Agoseris grandiflora var. *leptophylla* (Puget Sound agoseris), 60, 70, 136
agriculture, 30, 37, 46, 49, 54
Agrostis capillaris (colonial bentgrass), 23, 50, 65, 121, 135, 136
Agrostis gigantea [A. alba], 135
Agrostis pallens (seashore bentgrass), 135
Agrostis tenuis (colonial bentgrass), 23, 50, 65, 121, 135
Aira caryophyllea var. *caryophyllea* (silver hairgrass), 65, 121, 136
Aira praecox (little hairgrass), 65, 122, 136
Alaska oniongrass, 65, 127
Alnus rubra (red alder), 10
Alopecurus aequalis (shortawn foxtail), 135
Amelanchier alnifolia (serviceberry), 25, 63, 100, 133, 137
American bistort, 9, 24
American vetch, 62, 87, 132
Anaphalis margaritacea (western pearly everlasting), 60, 71, 137
Anemone lyallii (Lyall's anemone), 63, 97, 136
Angiosperms, 31, 60-65, 68-135
annual bluegrass, 50, 65, 129, 135
annual plants, 23-24, 41. *See also* specific plants
Antennaria howellii ssp. *howellii [Antennaria neglecta* var. *howellii]*, 60, 71, 130, 136
Antennaria neglecta var. *howellii* (field pussytoes), 60, 71, 130
Anthoxanthum odoratum (sweet vernalgrass), 23, 65, 122, 135, 137, 138
Anthriscus caucalis (burr chervil), 130
Anthriscus scandicina (burr chervil), 130

Aphanes australis (small-fruited parsley-piert), 133
Apiaceae [Umbelliferae], 9, 60, 68-69, 130, 144
Appendix A, 130-135
Appendix B, 136-138
Apocynaceae (dogbane family), 60, 70, 130
Apocynum androsaemifolium (spreading dogbane), 60, 70, 130, 136
Aquilegia formosa var. *formosa* (western columbine), 56, 63, 97, 137
Arabidopsis thaliana (thale cress), 131
Arabis eschscholtziana [Arabis hirsuta var. *eschscholtziana],* 131
Arctostaphylos uva-ursi (kinnikinnick), 61, 83, 132, 137
Armeria maritima ssp. *californica* (thrift, sea thrift, sea-pink), 62, 94, 133, 136
Arrhenatherum elatius (tall oatgrass), 23, *38,* 135, 137
Artemisia (sagebrush and/or wormwood), 9, 10
Artillery Impact Area, 6, 53-54
ash (volcanic), 8, 21, 39
Asian settlers, 47
Asteraceae [Compositae], 9, 60-61, 70-76, 130-131, 140, 141, 142
Aster curtus, 61, 75, 131, 146, 147
aster family, 9, 60-61, 70-76, 130-131, 140, 141, 142

B
bald-hip rose, 63, 103
Balsamorhiza deltoidea (deltoid balsamroot), ii, 22, 32, 60, 71, 130, 136
Barbarea orthoceras (yellow rocket), 131
barberry family, 61, 77, 131
barestem biscuitroot, 60, 68, 130
barestem teesdalia (shepherd's cress), 61, 80, 131
barren brome, 65, 123, 135
basketry materials, 70
bedstraw (cleavers), 24, 63, 104, 134
beech family, 62, 89
beef production, 45
Berberidaceae (barberry family), 61, 77, 131
Berberis aquifolium (tall Oregon-grape), 61, 77, 136
bigleaf lupine, 62, 85
biological methods of noxious weed control (biocontrol), 38-39
Bistorta bistortoides [Polygonum bistortoides], 9, 24
bittercress, 61, 78, 131, 143
black bentgrass, 135
black hawthorn, 63, 100
Black Hills, 13, 17

Black River, 5, 143
Black River-Mima Mounds Glacial Heritage Preserve (Glacial Heritage), ii, vii, 5, 30, 59
blacksnake root, 60, 69
bladder campion, 132
blanket flower, 60, 73
bluebell-of-Scotland, 61, 80
blue elderberry, 81
blue-eyed grass, 64, 115, 134
blue field madder, 63, 104, 134
blue-eyed Mary, 33, 45, 64, 108, 109
blue-lips blue-eyed Mary, 33, 64, 108, 109
blue scorpion-grass, 131
blue toadflax, 63, 110
blue wildrye, 25, 65, 135
Bombus californicus (bumblebee), 30
borage family, 61, 77, 131
Boraginaceae (borage family), 61, 77, 131
bracken fern, 10, 11, 60, 66
Brassicaceae [Cruciferae], 61, 78- 80, 131, 143
Bretz, J. Harlan, 16, 17, 18, 51
Brice, Cheryl, 58
British, 44-48
British Columbia, 29, 83, 87, 89, 92, 103, 110, 120
broad-leaved shooting star, 63, 96, 133
Brodiaea congesta (ookow, forktooth ookow, northern saitas), 64, 118
Brodiaea coronaria (crown brodiaea, harvest brodiaea), 64, 116, 135, 136
Brodiaea howellii (Howell's brodiaea), 64, 119
Brodiaea hyacinthina (white brodiaea), 64, 120
Bromus carinatus var. *carinatus* (California brome), 65, 123, 136
Bromus sterilis (barren brome, poverty brome), 65, 123, 135, 137
broomrape, 62, 93, 133 , 142
broomrape family, 62, 93, 133
bryophytes, 35, 37, 39, 40
buckthorn family, 63, 99, 133
buckwheat family, 62, 95, 133
Budd Inlet, 48
bull thistle, 130
bumblebee, 30, 92
bunchgrass, 22, 31, 49, 124, 125
burning: to favor oaks, 24, 57; golden paintbrush and, 30; to increase camas,

22, 43-44; to increase grasslands, 11, 21, 43-44, 57; Indigenous use of, (*see* Indigenous burning); streaked horned lark and, 35; wildlife and, 22, 35. *See also* fire suppression; prescribed burning
burr chervil, 130
burrowing clover, 62, 87, 132
buttercup, 22, 33, 63, 98-99, 133
buttercup family, 63, 97-99, 133
butterflies, 24, *28*, 29, 31-34, 37, 39, 40, 41, 54

C

California brome, 65, 123
California oatgrass, 22, 65, 123, 135
California poppy, 50
camas, ii, 22, 25, 32, 43-44, 45, 50, 55, 57, 58, 64, 98, 117, 135
camas consumption by Native Americans, 43, 55, 57, 117
Camassia leichtlinii var. *suksdorfii* (large camas), 32, 64, 117, 136
Camassia quamash (common camas), ii, 22, 25, 32, 33, 43-44, 45, 50, 55, 57, 58, 64, 98, 117, 135, 136
Camassia quamash ssp. *azurea,* 117, 135
Camassia quamash ssp. *maxima* (dark camas), 117, 135
Campanulaceae (harebell family), 61, 80, 131
Campanula rotundifolia (bluebell-of-Scotland, common harebell), 61, 80, 131, 136, 139
Caprifoliaceae (honeysuckle family), 61, 81-82, 141
Capsella bursa-pastoris (shepherd's purse), 50, 61, 78, 80, 131, 137, 143
Cardamine hirsuta (hairy bittercress), 131
Cardamine oligosperma (little western bittercress, shotweed), 61, 78, 131, 137, 143
Carex, 24, 25, 64, 114, 134, 142
Carex inops ssp. *inops* (long-stolon sedge), 25, 64, 114, 134, 136
Carex pensylvanica var. *vespertina* (long-stolon sedge), 25, 64, 114, 134
carrot family, 9, 60, 68-69, 130, 144
Caryophyllaceae (pink family), 61, 82, 131
Cascade Range (Cascades), 8, 12-17, 24, 30, 33, 43, 47, 48, 73, 74, 76, 83, 85, 87, 92, 93, 94, 95, 105, 109, 110, 111, 114, 119, 122
Cascadia Prairie-Oak Partnership, 33
cascara sagrada, 63, 99, 133
Castilleja hispida var. *hispida* (harsh paintbrush), 32, 63, 107, 134, 136
Castilleja levisecta (golden paintbrush), 24, *28*, 29, 30, 32, 63, 107, 136
cattle, 45, 46, 49
Cavness Ranch, 6

Ceanothus sanguineus (redstem ceanothus), 133
Center for Natural Lands Management (CNLM), vii, viii, 1, 5, 8, 20, 33, 35, 40, 55
Centralia, 1, 12, 16
Centaurium erythraea [C. umbellatum misapplied*]*, 132
Cerastium arvense ssp. *strictum* (field chickweed, mouse-ear), 32, 61, 82, 131, 136
Cerastium glomeratum (sticky mouse-ear chickweed), 132
Chamerion angustifolium (fireweed), 62, 92, 136
charcoal deposits, 10
Chatham Island, 58
Chehalis, 13, 14
Chehalis River, 9, *15*, 16, 24, 43, 45
Chehalis Tribe, 53
chickweed, 50, 132
chocolate lily, 64, 119, 120, 135, 143, 144
cinquefoil, *38*, 63, 102, 134
Cirsium vulgare (bull thistle), 130
citizen science, 2
Clarkia amoena var. *lindleyi* (farewell-to-spring), 62, 92, 136
clasping Venus'-looking-glass, 61, 81
Claytonia perfoliata (miner's lettuce), 62, 95, 133, 136, 143
cleavers, 63, 104, 134
climate, 1, 7-12, 21, 43, 55
climate change: in the early Holocene, 7, 10-12, 21, 55; and Indigenous burning, 55; in the present (late Holocene), 11-12, 43; and vegetation change, 8-11, 21-22
Clinopodium douglasii (yerba buena), 62, 91, 136
Clusiaceae [Hypericaceae], 62, 90, 132
CNLM (Center for Natural Lands Management), vii, viii, 1, 5, 8, 20, 33, 35, 40, 55
coastal manroot, 61, 82, 132
coast goldenrod, 61, 75, 131
Collinsia, 33, 45, 64, 108, 109, 138
Collinsia grandiflora (large-flowered blue-eyed Mary, giant blue-eyed Mary, blue-lips blue-eyed Mary), 33, 64, 108, 109, 138
Collinsia parviflora (small-flowered blue-eyed Mary, maiden blue-eyed Mary), 33, 64, 108, 109, 138
colonial bentgrass, 23, 50, 65, 121, 135
colonization, 12, 25, 55
Columbia River, 31, 43, 47, 48, 92

columbine, 56, 63, 97
common centaurium, 132
common dandelion, 32, 61, 70, 74, 76, 131, 139, 143
common gaillardia, 60, 73
common harebell, 61, 80, 131, 139
common panicgrass (witchgrass), 65, 127, 134, 135
common plantain, 32, 33, 62, 94, 133 , 139
common snowberry, 25, 61, 81
common St. John's-wort, 23, 50, 62, 90, 132
common stork's-bill, 132
common tansy, 131
common teasel, *38*
common velvet grass, 23, 50, 65, 125, 135
common vetch, 33, 62, 88, 132, 142
common viburnum, 61, 82
common woolly sunflower, 22-23, 29, 60, 72, 130, 142
common yampah, 130
community-based learning, 1. *See also* volunteers
competitive release, 25
Compositae, 9, 60-61, 70-76, 130-131, 140, 141, 142
congested snake lily, 64, 118
conifers, 1, 3, 12, 21, 22, 24, 27, 51, 60, 66-67, 81, 89, 130, 140. *See also*
    Douglas-fir; hemlock; pines; western red cedar
Consortium of Pacific Northwest Herbaria, 2
Cooper, James Graham, 43-45, 47, 50, 55,
Cordilleran Ice Sheet, 8, 14, 21
corn mint, 133
corn speedwell, 64, 111
Cowlitz River, 45
Cowlitz Trail, 47
Crataegus douglasii (black hawthorn), 63, 100, 136
Crataegus monogyna (one-seed hawthorn), 63, 101, 136
Crater Lake, 8
creeping yellow wood-sorrel, 133
Crepis capillaris (smooth hawksbeard), 60, 72, 130, 137
Crockett, Walter, 47
crown brodiaea (harvest brodiaea), 64, 116, 135
Cruciferae (mustard family), 61, 78- 80, 131, 143
cucumber family, 61, 82, 132
Cucurbitaceae (cucumber family), 82, 132
cut-leaf geranium, 132

cutleaf microseris, 61, 74, 131
Cynoglossum officinale (common hound's-tongue), 131
Cyperaceae (sedge family), 9, 64, 114, 134
Cytisus scoparius (Scotch broom, Scot's broom), 23, 33, 35, 36, 37, *38*, 39, 40, 50, 55, 61, 75, 83, 132, 137, 142

D

dairy production, 46, 48
dandelion, 32, 61, 70, 74, 76, 131, 139, 143
Danthonia californica (California oatgrass), 22, 65, 123, 135, 137
Danthonia intermedia (timber oatgrass), 65, 124, 137
Danthonia spicata (poverty oatgrass), 135
Daphne laureola (laurel spurge), *38*
dark camas, 135
dark-throated shooting star, 63, 96
Daucus carota (wild carrot), 60, 68, 138
dead-nettle, 62, 91, 132
death camas, 64, 120, 135
deer: forage improved with burning, 43
deer vetch, 61, 83, 132
Delphinium nuttallianum (upland larkspur), 63, 98, 137
deltoid balsamroot, ii, 22, 32, 60, 71, 130
Deschampsia cespitosa (tufted hairgrass), 24, 29
Deschutes Prairie, *6*
dewberry (trailing blackberry), 63, 103, 134
Dianthus deltoides (maiden pink), 132
Dichanthelium acuminatum ssp. *fasciculatum* (hairy panicgrass), 65, 128, 135
Dichelostemma congestum (ookow, forktooth ookow, northern saitas, congested snake lily), 64, 118, 138
Dicotyledoneae, 60-64, 68-113, 130-134
Dipsacus fullonum (common teasel), *38*
Discovery Island, 58
Dodecatheon hendersonii (Henderson's shooting star, broad-leaved shooting star), 63, 96, 133, 136
Dodecatheon pulchellum var. *pulchellum* (dark-throated shooting star, sticky shooting star), 63, 96, 136
dogbane family, 60, 70, 130
Donation Land Act, 49
Donation Land Law of Washington, 47-48
Douglas-fir, 9, 10, 22, 23, 25, 31, 44, 51, 53, 55, 60, 67, 75, 130, 140
dovefoot geranium, 62, 90 132

Draba verna (spring draba, spring whitlow-grass), 61, 79, 131, 137, 139
Dunn, Patrick, viii, 55
dwarf owl-clover, 33, 64, 111, 134

E
early blue violet, iv, 33, 64, 113, 134
ecological nostalgia, 57
ecological restoration, 3, 24, 57-58. *See also* restoration ecology
elk, 50
Elymus glaucus ssp. *glaucus* (blue wildrye), 25, 65, 135, 136
endangered habitat, 23, 54
endangered landscape, 36
endangered species, 28, 29, 32, 33, 34, 37
Endangered Species Act, 3, 27, 35, 54
English plantain (narrowleaf plantain), 62, 93, 133
Epilobium angustifolium, 62, 92
Eremophila alpestris strigata (streaked horned lark), 24, 28, 31, 34-35, 37, 40, 54
Ericaceae (heath family), 61, 83, 132
Erigeron speciosus (showy fleabane), ii, 60, 72, 130, 138
Eriophyllum lanatum var. *leucophyllum* (Oregon sunshine, common woolly sunflower), 22-23, 29, 60, 72, 130, 136, 142
Erodium cicutarium (common stork's-bill), 132
Erythronium oregonum (white fawn lily), 64, 118, 138
Eschscholtzia californica (California poppy), 50
Euphorbia esula (leafy spurge), 38
Euphydryas editha ssp. *taylori* (Taylor's checkerspot butterfly), 24, 28, 31, 32-33, 37, 54
Euro-American settlers, 12, 21, 25, 27, 43- 55, 57
European centaurium, 132
evening primrose family, 62, 92
Evergreen Herbarium, vii, viii, 1-2, 59, 130
Evergreen Natural History Museum, vii, viii, 59
exotic plants, 23, 31, 37, 38, 49, 50, 51, 55. *See also* specific taxa

F
Fabaceae [Leguminosae], 61, 83-88, 132, 139, 142
Fagaceae (beech family), 62, 89
farewell-to-spring, 62, 92
fawn lily, 64, 118
fern, 10, 11, 60, 66, 140, 143

Festuca idahoensis var. *roemeri* (Idaho fescue), 22, 65, 125, 135
Festuca occidentalis (western fescue), 135
Festuca roemeri (Roemer's fescue), 22, 65, 125, 135, 136
Festuca rubra (red fescue), 29, 135
field chickweed (mouse-ear), 32, 61, 82, 131
field cress, 61, 79, 131
field pepperweed, 61, 79, 131
field pussytoes, 60, 71, 130
figwort family, 32, 63, 107-111, 134
Filago minima (little cottonrose), 130
fir, 8-9
fire: *See* burning; fire suppression; prescribe burning
fire suppression, 30, 51, 57
fireweed, 62, 92
flowering plants, 31, 60-65, 68-135
focal restoration, 58
food systems: changes in Indigenous, 3, 5, 30, 46, 48, 49, 50, 53; renewed interest in traditional Indigenous, 3
forktooth ookow, 64, 118
Fort Lewis, ii, vii, 30, 53, 54
Fort Nisqually, 45, 48
Fort Vancouver, 45, 48
fossil: macrofossils, 7-8; microfossils (pollen), 7-11, 12
Fragaria vesca (woods strawberry), 133
Fragaria virginiana (wild strawberry), 32, 63, 101, 134, 136
Frangula purshiana (cascara sagrada), 63, 99, 133, 136
Frasier Glaciation, 7
Friends of Puget Prairies (FOPP), 2
fringe cup, 63, 106
Fritillaria affinis (chocolate lily), 64, 119, 120, 135, 137, 143, 144
Fritillaria lanceolata (chocolate lily), 64, 119, 120, 135
fur traders, 44, 45

G

Gaillardia aristata (blanket flower, common gaillardia), 60, 73, 136
Galium aparine (cleavers, bedstraw, goosegrass), 63, 104, 134, 136
Galium boreale (northern bedstraw), 24
Galium parisiense (wall bedstraw), 63, 104, 136
game, 43
Garry oak, 10, 23, 62, 89, 120
geology of south Puget Sound prairies, 7, 13-20

Geraniaceae (geranium family), 62, 90, 132
Geranium dissectum (cut-leaf geranium), 132
Geranium molle (dovefoot geranium), 62, 90, 132, 136
giant blue-eyed Mary, 33, 64, 108, 109
Glacial Heritage, ii, vii, 5, 30, 59
glacial lakes, 15, 16
glacial outburst flood (jökulhlaup), 13, 15, 18
glacial outwash, 1, 9, 17, 18, 21, 22-25
glaciation, 7, 8, 9, 14, 16, 21, 53
Glacier Peak, 8
Gnaphalium palustre (lowland cudweed), 130
golden paintbrush, 24, *28*, 29, 30, 32, 63, 107
goldenrod, 61, 75, 130
goosegrass, 63, 104, 134
Gramineae (grass family), 65, 121-129, 135
gravel mining, 53
gravelly outwash plains, 17. *See also* glacial outwash
Great Lakes, 52
great plantain (common plantain), 32, 33, 62, 94, 132, 139

H
habitat fragmentation (fragmentation), 36, 37, 53, 55
hairgrass, 24, 29, 65, 121, 122, 135
hairy bittercress, 131
hairy cat's ear, 23, *38*, 61, 74, 76, 130
hairy honeysuckle, 25
hairy panicgrass, 65, 128, 135
hairy plantain, 62, 94
hairy vetch, 62, 88, 132
Hall's aster, *28*, 31
hand-pulling of weeds, 38-39
Hansen, Henry, 8
harebell family, 61, 80
harsh paintbrush, 134
harvest brodiaea, 64, 116, 135
Hawaiians, 47
hawthorn, 63, 100, 101
heal-all (self-heal), 62, 92, 133,
heath family, 61, 83, 132
hemlock, 9-11
Henderson's shooting star (broad-leaved shooting star), 63, 96, 133

herbarium specimens of prairie plants, vii, viii, 1, 2, 23, 29, 31, 59, 130-135. *See also* Appendix A
herbicide application, 38, 39
Heuchera micrantha (smallflower alumroot), 134
Hieracium albiflorum (white-flowered hawkweed), 60, 73, 130, 136
Hieracium pilosella (mouse-eared hawkweed), *38*
Hieracium scouleri (Scouler's hawkweed), 60, 73, 136
Higgs, Eric, 57-58
Holcus lanatus (common velvet grass), 23, 50, 65, 125, 135, 136
Holocene, 10, 11
Homestead Act of 1862, 48
honeysuckle family, 61, 81-82, 141
hooded ladies'-tresses, 65, 121
Howell's brodiaea, 64, 119, 136
Howell's violet, 134
Hudson's Bay Company, 45, 47, 48
Hypericaceae (St. John's-wort family), 62, 90, 132
Hypericum perforatum (common St. John's-wort), 23, 50, 62, 90, 132, 137
Hypochaeris radicata (hairy cat's ear), 23, *38*, 61, 74, 76, 130, 137
Hypsithermal, 22

I
illustrators for this project, 136-138
Indian plum, 63, 102
Indian wheat, 62, 94
Indigenous burning: for camas, 22, 43-44; to foster prairies, 1, 11, 22; by the Salish, 45, 51, 54, 55
Indigenous Peoples' Restoration Network, 58
Indigenous women, 17
insect pollinators, 29, 30, 31-32, 49
introduced plants (exotic plants), 23, 31, 37, *38*, 49, 50, 51, 55. *See also* specific taxa
Iridaceae (iris family), 64, 114-115, 134
Iris tenax (Oregon iris, tough-leaf iris), 64, 114, 136, 140
Italian ryegrass, 65, 126, 135

J
Johnson Prairie, 5, 6
Joint Base Lewis-McChord (JBLM), viii, 5, 24, 41, 54, 67. *See also* Fort Lewis
Juan de Fuca plate, 13
Juncaceae (rush family), 64, 115, 134

Juncus bufonius var. *bufonius* (toad rush), 64, 115, 136

K
Kentucky bluegrass, 65, 129, 135
kinnikinnick, 61, 83, 132
Koeleria cristata (prairie Junegrass), 65, 125, 135, 137
Koeleria macrantha (prairie Junegrass), 65, 125, 135, 137

L
Labiatae (mint family), 62, 91-92, 133
Lake Bretz, 16
lamb's lettuce, 134
Lamiaceae [Labiatae], 62, 91-92, 133
Lamium purpureum (dead-nettle), 62, 91, 133, 136
land enclosure, 47, 48
large camas, 32, 64, 117
large-flowered agoseris (Puget Sound agoseris), 60, 70
large-flowered blue-eyed Mary, 33, 64, 108, 109
large-leaved lupine, 62, 85
larkspur, 63, 98, 143
laurel spurge, *38*
leafy spurge, *38*
least hop clover, 62, 86
Leguminosae (pea family), 61, 83-88, 132, 139, 142
Lepidium campestre (field pepperweed, pepperwort, field cress), 61, 79, 131, 137
Leucanthemum vulgare (oxeye daisy), 23, *38*, 50, 61, 74, 137, 141
licorice fern, 60, 66, 140, 143
lidar, 17, 19, 20
Liliaceae (lily family), 64, 116-120, 135
Lilium columbianum (tiger lily), 64, 119, 135, 137
Linaria canadensis (blue toadflax), 63, 110
list of illustrators (Appendix B), 136-138
list of herbarium voucher specimens (Appendix A), 130-135
Lithophragma parviflorum (small-flower woodland star, small-flower prairie star), 63, 105, 136
little buttercup, 63, 99
little cottonrose, 130
little hairgrass, 65, 122, 135
little western bittercress, 61, 78, 131, 143
livestock grazing, 5, 30, 48, 49, 50, 53. *See also* beef production; cattle; pigs;

sheep
lodgepole pine, 9-10
Logfia minima (little cottonrose), 130
logging, 53
Lolium multiflorum (Italian ryegrass), 65, 126, 135, 138
Lolium perenne (perennial ryegrass), 65, 126, 135, 137
Lomatium nudicaule (barestem biscuitroot, pestle parsnip), 60, 68, 130, 136
Lomatium triternatum (nineleaf biscuitroot), 60, 68, 130, 137
Lomatium utriculatum (spring gold), 23, 33, 60, 69, 130, 138
long-stolon sedge, 25, 64, 114, 134
Lonicera hispidula (hairy honeysuckle), 25
Lotus micranthus (short-flower bird's-foot-trefoil, small-flowered deer vetch), 61, 83, 132
Lotus nevadensis var. *douglasii* (Nevada deervetch), 132
Lower Weir Prairie, 5, 6
lowland cudweed, 130
Lupinus albicaulis (sicklekeel lupine, white-stemmed lupine), 23, 62, 84, 137
Lupinus bicolor (two-color lupine, miniature lupine, small-flower lupine), 62, 84, 132, 137
Lupinus lepidus (prairie lupine), 33, 62, 85, 132, 136
Lupinus polyphyllus var. *pallidipes* (bigleaf lupine, large-leaved lupine), 62, 85, 137
Luzula comosa var. *laxa* (Pacific woodrush), 64, 116, 134, 136, 140
Luzula subsessilis (short-stalked wood-rush), 134
Lyall's anemone, 63, 97

M

madder family, 63, 104, 134
maiden blue-eyed Mary, 33, 64, 108, 109
maiden pink, 132
map of south Puget Sound prairies, 6
Marah oregana (coastal manroot, wild cucumber), 61, 82, 132, 138
Mardon skipper butterfly, 24, *28*, 31, 33-34
Marion Prairie, 5
marsh speedwell, 32
Martinez, Dennis, 58
Mazama pocket gopher, 21, 24, *28*, 31, 35-36, 37, 40, 54
meadow death-camas, 64, 120, 135
Melica subulata var. *subulata* (Alaska oniongrass), 65, 127, 137
Mentha arvensis (corn mint), 133
Micranthes integrifolia (swamp saxifrage, whole-leaf saxifrage), 63, 106, 134,

136
Microseris laciniata ssp. *laciniata* (cutleaf microseris), 61, 74, 131, 137
Microsteris gracilis (slender phlox), 133
military training, 53
Mima Mounds (and Mima mounds), 5, 6, 15, 17, 18, 19, 25, 30, 129
Mima Prairie, *18*
Mineral Lake, 11
miner's lettuce, 62, 95, 133, 143
miniature lupine (two-color lupine), 62, 84, 132
mint family, 62, 91-92, 133
Mississippi River, 43
Missouri goldenrod, 61, 75, 131
Monocotyledoneae, 64-65, 114-129, 134-135
Montesano, 52
Montia perfoliata (miner's lettuce), 62, 95, 133, 143
moss, 35, 37, 39, 40
mountain hemlock, 9-10
Mount Albert goldenrod, 61, 75, 131
Mount Mazama, 8
Mount Rainier, 16, 17
Mount St. Helens, 8
mouse-ear (field chickweed), 32, 61, 82, 131
mouse-eared hawkweed, *38*
mowing, 38-39
mule's ears, 61, 76
mustard family, 61, 78- 80, 131, 143
mutton production, 45
mycorrhizal fungi, 40
Myosotis discolor (yellow and blue forget-me-not), 61, 77, 131, 138
Myosotis stricta [M. micrantha], 131

N
naked broomrape, 62, 93, 133, 142
narrowleaf plantain, 2, 93, 133
narrow-leaved mule's ears, 61, 76
Natural Area Preserves, 5
Natural Resources Conservation Service (NRCS), 21
Nature Conservancy (TNC), viii, 5, 6
nettle, 64, 112
nettle family, 64, 112
Nevada deervetch, 132

nineleaf biscuitroot, 60, 68, 130
nippleseed (common plantain), 32, 33, 62, 94, 133, 139
Nisqually Plain, 46
Nisqually River, *15*, 45
Nisqually Tribe, 45, 53
nitrogen, 23, 39, 40, 55
nitrogen-fixing bacteria, 23, 55
nitrogen-poor soils, 23, 40
nomenclature, 1, 59
non-native grasses, 22, 40
North American plate, 13
Northern Pacific Railroad, 52
northern saitas, 64, 118
Norton, Helen, 46
Nuttallanthus texanus [Linaria canadensis], 63, 110, 137

O

oak, 10, 23, 62, 89, 120. *See also* oak savanna and oak woodland
oak savanna, 24-25, 53, 57
Oakville, 12
oak woodland, 1, 2, 3, 25, 57, 51, 59, 81, 89, 105, 118, 119
Oemleria cerasiformis (Indian plum, osoberry), 63, 102, 138
Olympia, i, vii, viii, 7, 8, 12, *15*, 16, 17, 18, 24, 34, 52
Olympic Mountains, 12, 13, 14, 16, 17
Onagraceae (evening primrose family), 62, 92
one-seed hawthorn, 63, 101
ookow, 64, 118
Oregon, 24, 29, 31, 34, 35, 37, 46, 74, 75, 80, 112
Oregon ash, 24, 143
Oregon Department of Fish and Wildlife, 35
Oregon iris, 64, 114, 140
Oregon spotted frog, 24
Oregon State University, 35
Oregon sunshine, 22-23, 29, 60, 72, 130, 142
Oregon Trail, 46
Oregon white oak (Garry oak), 10, 23, 62, 89, 120
Oregon zoo, 33
Orchidaceae (orchid family), 65, 121
Orobanchaceae (broomrape family), 62, 93, 133, 135
Orobanche uniflora (naked broomrape), 62, 93, 133, 136, 142
Orthocarpus pusillus (dwarf owl-clover), 33, 64, 111, 134

osoberry, 63, 102
outwash plains, 17. *See also* glacial outwash
oval-leaved viburnum, 61, 82
Oxalis corniculata (creeping yellow wood-sorrel), 133
Oxalidaceae (wood-sorrel family), 133
oxeye daisy, 23, *38*, 50, 61, 74, 141

P

Pacific coast rockcress, 131
Pacific Northwest (PNW), 1, 2, 8, 10, 11, 12, 21, 27, 29, 31, 37, 46, 53, 59, 70, 86, 105
Pacific Ocean, 7, 9, 11, 43, 52
Pacific Plate, 13
Pacific sanicle, 60, 69
Pacific woodrush, 64, 116, 134, 140
Panicum capillare ssp. *capillare* (witchgrass, common panicgrass), 65, 127, 135, 136
Panicum occidentale (hairy panicgrass), 65, 128, 135, 136
Parentucellia viscosa (yellow glandweed), 64, 110, 134, 137
pasture grasses, 23, 37, 51, 121, 122. *See also* specific taxa
pea family, 61, 83-88, 132, 139, 142
pearly everlasting, 60, 71
pepperwort (field pepperweed), 61, 79, 131
perennial ryegrass, 65, 126, 135
Perideridia gairdneri ssp. *borealis* (common yampah), 130
pestle parsnip, 60, 68, 130
Phalaris arundinacea (reed canary grass), *38*
Phleum pratense (Timothy), 50, 65, 128, 135, 136, 137
phlox family, 133
Picea sitchensis (Sitka spruce), 9-11
Pierce County, 24, 53
pigs, 50
Pinaceae (pine family), 66-67, 130
pine bluegrass, 22
pines, 8, 9, 11, 24, 31, 60, 66, 67, 130
pink family, 61, 82, 131
Pinus contorta var. *contorta* (shore pine), 60, 66, 130, 137, 138
Pinus contorta var. *latifolia* (lodgepole pine), 9
Pinus monticola (western white pine), 11
Pinus ponderosa var. *ponderosa* (ponderosa pine), 24, 31, 60, 67, 136
Plagiobothrys scouleri var. *scouleri* (Scouler's popcorn flower), 61, 77, 137

Plantaginaceae (plantain family), 62, 94, 133
Plantago lanceolata (narrowleaf plantain, English plantain), 2, 93, 133, 137
Plantago major (common plantain, great plantain, nippleseed), 32, 33, 62, 94, 133, 137, 139
Plantago patagonica (hairy plantain, Indian wheat), 62, 94, 137
plantain family, 62, 94, 133
Plectritis congesta (seablush, rosy plectritis), 64, 112, 137
Plumbaginaceae (plumbago family), 62, 94, 133
Poa annua (annual bluegrass), 50, 65, 129, 135, 136
Poaceae [Gramineae] 65, 121-129, 135
Poa pratensis ssp. *pratensis* (Kentucky bluegrass), 65, 129, 135, 136
Poa secunda (pine bluegrass), 22
Polemoniaceae, 133
Polites mardon (Mardon skipper), 24, *28*, 31, 33-34
pollen record, 7-11. *See also* pollination
pollination, 8, 29, 32, 37, 84
pollinators, 29, 30, 32, 49
Polygonaceae (buckwheat family), 62, 95, 133
Polygonum bistortoides (American bistort), 9, 24
Polypodiaceae (polypoidy family), 60, 66
Polypodium glycyrrhiza, 60, 66, 138, 140, 143
polypody family, 60, 66
ponderosa pine, 24, 31, 60, 67
popcorn flower, 61, 77
pork production, 50
Portulacaceae (purslane family), 62, 95, 133
postglacial vegetation, 8-11
potato, 46, 48
Potentilla gracilis var. *gracilis* (slender cinquefoil), 63, 102, 134, 137
Potentilla recta (sulphur cinquefoil), *38*
poverty brome, 65, 123, 135
prairie Junegrass, 65, 125, 135
prairie lupine, 33, 62, 85, 132
precipitation, 10, 12, 22
Preemption Act of 1841, 46, 48
prescribed burning, 24, *38*, 39, 54
prickly sow thistle, 50
Primulaceae (primrose family), 63, 96, 133
Prunella vulgaris (self-heal, heal-all), 62, 92, 133, 137
Pseudotsuga menziesii var. *menziesii*, 9, 10, 22, 23, 25, 31, 44, 51, 53, 55, 60, 67, 75, 130, 136, 140

Pteridium aquilinum var. *pubescens* (bracken fern), 10, 11, 60, 66, 137
Puget Lobe of the Vashon ice sheet, 7, 8, 14, 15, 16, 21
Puget Sound agoseris, 60, 70
Puget Sound Agricultural Company, 46, 49
Puget Trough, 7, 8, 10, 11, 12, 29
purslane family, 62, 95, 133
pussytoes, 60, 71, 130
Puyallup River, 45

Q

Queen Anne's-lace, 60, 68
Quercus garryana var. *garryana* (Oregon white oak), 10, 23, 62, 89, 120, 136, 137

R

railroad, 52-53
railroad survey, 43, 50
Rana pretiosa (Oregon spotted frog), 24
Ranunculaceae (buttercup family), 63, 97-99, 133
Ranunculus, 22, 33, 45, 63, 98-99, 133, 137, 138
Ranunculus occidentalis var. *occidentalis* (western buttercup), 22, 33, 63, 98, 99, 133, 138
Ranunculus uncinatus (little buttercup), 63, 98, 99, 137
rare plants and animals, 27, *28*. See also sensitive species
red alder, 10
red clover, 62, 86, 132
red elderberry, 61, 81
red fescue, 29, 135
red sandspurry, 50, 132
red-seeded dandelion, 131
redstem ceanothus, 133
reed canary grass, *38*
restoration ecology, 1, 37-41. *See also* ecological restoration
Rhamnaceae (buckthorn family), 63, 99, 133
Rhamnus purshiana (cascara sagrada), 63, 99, 133
riparian vegetation (riparian woodland), 25, 105, 106, 143
Rocky Mountains Canada goldenrod, 131
Rocky Mountain subalpine zone, 9
Rocky Prairie, 5, 6, 16, *19*, 36
rodent, 17, 21, 24, *28*, 29, 31, 35-36, 37, 40, 54
Roemer's fescue, 22, 65, 125, 135

Rosaceae (rose family), 100-103, 133
Rosa gymnocarpa (bald-hip rose), 63, 103, 137
rosy plectritis, 64, 112
Rubiaceae (madder family), 63, 104, 134
Rubus spectabilis (salmonberry), 63, 103, 137
Rubus ursinus (trailing blackberry, dewberry), 63, 103, 134, 137
Rumex acetosella (sheep sorrel), 62, 95, 133, 138
rush family, 64, 115
Russian American Company, 45

S

sagebrush-steppe, 68, 112, 120
sagebrush/wormwood, 9, 10
Sagina decumbens ssp. *occidentalis* (western pearlwort), 132
Salish People, 44, 45, 48, 49, 51, 54, 55, 66
Salish Sea, 2, 32, 45
Salicaceae (willow family), 63, 105
Salix (willow), 10, 63, 105
Salix scouleriana (Scouler's willow), 63, 105, 136
salmonberry, 63, 103
Sambucus nigra ssp. *caerulea* (blue elderberry), 81
Sambucus racemosa var. *racemosa* (red elderberry), 61, 81, 136
Sanicula crassicaulis (Pacific sanicle, blacksnake root), 60, 69, 136
Satureja douglasii, 62, 91
Saxifragaceae (saxifrage family), 63, 105-106, 134
Saxifraga integrifolia (swamp saxifrage, wholeleaf saxifrage), 63, 106, 134
Scatter Creek Wildlife Area, 5, 24, 26, 34, 59
Sciurus griseus (western gray squirrel), 25, 67
Sciurus Wildlife Area, 34
Scotch broom (Scot's broom), 23, 33, 35, 36, 37, 38, 39, 40, 50, 55, 61, 75, 83, 132, 142
Scouler's hawkweed, 60, 73
Scouler's popcorn flower, 61, 77
Scouler's silene, 132
Scouler's willow, 63, 105
Scrophulariaceae (figwort family), 32, 63, 107-111, 134, 135
seablush, 64, 112
sea-pink (sea thrift), 62, 94, 133
seashore bentgrass, 135
sea thrift, 62, 94, 133
Seattle, 14, 15, 52

sedge, 9, 24, 25, 64, 114, 134
sedge family, 9, 64, 114, 134
self-heal (heal-all), 62, 92, 133
Senecio jacobaea (tansy ragwort), *38*, 131
Senecio sylvaticus (wood groundsel), 131
sensitive species, 27-36. See also endangered species
Sericocarpus rigidus (white-topped aster), 28, 30-31, 35, 61, 75, 131, 136
SER's Indigenous Peoples' Restoration Network, 58
serviceberry, 25, 63, 100, 133
sheep, 45, 46, 49
sheep sorrel, 62, 95, 133
shepherd's cress (barestem teesdalia), 61, 80, 131
shepherd's purse, 50, 61, 78, 80, 131, 143
Sherardia arvensis (blue field madder), 63, 104, 134, 137
shooting star, 63, 96, 133
shore pine, 9, 60, 66, 130
shortawn foxtail, 135
short-stalked wood-rush, 134
shotweed (little western bittercress), 61, 78, 131, 143
showy fleabane, ii, 60, 72, 130
shrub-steppe, 22, 68, 79, 112, 120
sicklekeel lupine, 23, 62, 84
Silene scouleri ssp. *scouleri* (Scouler's silene), 132
Silene vulgaris [*S. cucubalus*], 132
silver hairgrass, 65, 121, 135
Sisyrinchium angustifolium, 134
Sisyrinchium idahoense (blue-eyed grass), 64, 115, 134, 137
Sisyrinchium idahoense var. *occidentale* (blue-eyed grass), 115
Sisyrinchium idahoense var. *segetum* (blue-eyed grass), 115
Sitka spruce, 9-11
slender cinquefoil, 63, 102, 134, 137
slender phlox, 133
smallflower alumroot, 134
small-flowered blue-eyed Mary, 33, 64, 108, 109
small-flowered deer vetch, 61, 83, 132
small-flowered trillium, 24, 64, 120, 135
small-flower lupine (two-color lupine), 62, 84, 132
small-flower prairie star, 63, 105
small-flower woodland star, 63, 105
small-fruited parsley-piert, 133
smallpox, 44

smooth hawksbeard, 60, 72, 130
snowberry, 25, 61, 81
Snowden, Clinton A., 46
snow-queen, 134
Society for Ecological Restoration (SER), 57
soils, 7, 9, 10, 17-21, 23, 24, 35-40, 43, 47, 49-50, 53-55
soil maps, 17, 18, 20
Solidago lepida var. *salebrosa* [*Solidago canadensis* var. *salebrosa*], 131
Solidago missouriensis var. *tolmieana* (Missouri goldenrod), 61, 75, 131, 136
Solidago simplex var. *simplex* [*Solidago spathulata* var. *neomexicana*], 61, 75, 131, 137
Solidago spathulata var. *neomexicana* (Mount Albert goldenrod, coast goldenrod), 61, 75, 131
Sonchus asper (prickly sow thistle), 50
Sonchus oleraceus (sow thistle), 50
Songhees First Nation, 58
South Sound Prairies Program, 55
South Weir Prairie, 5, 6, *6*
sow thistle, 50
speedwell, 32, 64, 111
Spergularia rubra (red sandspurry), 50, 132
Spiranthes romanzoffiana (hooded ladies'-tresses), 65, 121, 137
spreading dogbane, 60, 70, 130
spring draba, 61, 79, 131, 139
spring gold, 23, 33, 60, 69, 130
spring whitlow-grass (spring draba), 61, 79, 131, 139
spruce, 9-11
squirrels, 25, 67
Stellaria media (chickweed), 50, 132
Stevens, Isaac, 43, 47, 48, 52
sticky mouse-ear chickweed, 132
sticky shooting star, 63, 96
stinging nettle, 64, 112
St. John's-wort, 23, 50, 62, 90, 132
St. John's-wort family, 62, 90, 132
Strait of Juan de Fuca, 9, 13
streaked horned lark, 24, *28*, 31, 34-35, 37, 40, 54
stream violet, 64, 113, 134
study area description, 5-6
suckling clover, 62, 86
Sustainability in Prisons Project, 33, 41

swamp saxifrage, 63, 106, 134
sweet vernalgrass, 23, 65, 122, 135
Symphoricarpos albus var. *albus* (common snowberry), 25, 61, 81, 138
Symphyotrichum hallii (Hall's aster), *28*, 31
Synthyris reniformis (snow-queen), 134

T
Tacoma, 1, 13, 52, 53
Tacoma Fault, 13, 14
Tacoma, Olympia, and Gray's Harbor Railroad, 52
tall oatgrass, 23, *38*, 135
tall Oregon-grape, 61, 77
Tanacetum vulgare (common tansy), 131
tansy ragwort, *38*, 131
Tanwax-Ohop Creek, *14, 15*, 17, 18
Taraxacum erythrospermum (red-seeded dandelion), 131
Taraxacum officinale (common dandelion), 32, 61, 70, 74, 76, 131, 137, 139, 143
Taylor's checkerspot, 24, 28, 31, 32-33, 37, 54
tectonic activity, 13, 14, 17
Teesdalia nudicaulis (barestem teesdalia, shepherd's cress), 61, 80, 131, 137
Tellima grandiflora (fringe cup), 63, 106, 137
Tenalquot Prairie, 6, 30
Tenino, 8, 52
thale cress, 131
Thomomys mazama (mazama pocket gopher), 21, 24, *28*, 31, 35-36, 37, 40, 54
thrift, 62, 94, 133
Thuja plicata (western red cedar), 11
Thurston County, viii, 5, 16, 48, 53, 113
tiger lily, 64, 119, 135
Timothy, 50, 65, 128, 135
toad rush, 64, 115
Tolmie, William, 45, 46
tough-leaf iris, 64, 114, 140
Toxicoscordion venenosum [Zigadenus venenosus], 64, 120, 135
tracked and Stryker vehicles, 53
traditional ecological knowledge, 1. *See also* Indigenous burning
Tragopogon dubius (yellow salsify), 61, 76, 131, 136
trailing blackberry (dewberry), 63, 103, 134
treaties, 47, 52
Treaty of Medicine Creek, 52

Trifolium dubium (least hop clover, suckling clover), 62, 86, 137
Trifolium pratense (red clover), 62, 86, 132, 137
Trifolium subterraneum (burrowing clover), 62, 87, 132, 137
Trillium parviflorum (small-flowered trillium), 24, 64, 120, 136
Triodanis perfoliata (clasping Venus'-looking-glass), 61, 81, 137
Triphysaria pusilla (dwarf owl-clover), 33, 64, 111, 134, 136
Triteleia grandiflora (Howell's brodiaea), 64, 119, 137
Triteleia hyacinthina (white brodiaea), 64, 120, 136
Tsuga heterophylla (western hemlock), 9-11
Tsuga mertensiana (mountain hemlock), 9-10
tufted hairgrass, 24, 29
Tumwater (New Market), 47
tundra, 9
two-color lupine, 62, 84, 132

U
Umbelliferae (carrot family), 9, 60, 68-69, 130, 144
upland larkspur, 63, 98
upland yellow violet, 64, 113, 134
Upper Weir Prairie, 5, 6
U.S. Department of Defense (DoD), viii, 5, 33, 35, 54
U.S. Fish and Wildlife Service (USFWS), 27, 33, 35, 37
urbanization, 48, 53, 55
Urticaceae (nettle family), 64, 112
Urtica dioica ssp. *gracilis* (stinging nettle), 64, 112, 137

V
Valerianaceae (valerian family), 64, 112, 134
Valerianella locusta (lamb's lettuce), 134
Vancouver, 43
Vancouver, George, 44
Vancouveria hexandra (white inside-out-flower), 131
Vancouver Island, 30, 32, 43, 50, 58
Vashon glacier, 7, 8, 14, 15, 16, 21
vegetation, 1: descriptions, 21-26; historic, 45; studies based on fossilized pollen, 7-12; streaked horned lark and, 34; study at Mima Mounds, 129. *See also* climate change and vegetation change, fire suppression, oak savanna, oak woodland, riparian vegetation, sagebrush-steppe, tundra
vegetation management, 5, 6, 37-41, 57-58. *See also* burning
velvet grass, 23, 50, 65, 125, 135
Veronica arvensis (corn speedwell), 64, 111, 134, 137

Veronica beccabunga (American brooklime), 32
Veronica scutellata (marsh speedwell), 32
Veronica serpyllifolia ssp. *serpyllifolia* (thyme-leaved speedwell), 32
vetch, 33, 62, 87, 88, 132, 142
Viburnum ellipticum (common viburnum, oval-leaved viburnum), 61, 82, 136
Vicia americana var. *americana* (American vetch), 62, 87, 132, 137
Vicia hirsuta (hairy vetch), 132
Vicia sativa var. *angustifolia* (common vetch), 33, 62, 88, 132, 142
Vicia sativa var. *sativa* (common vetch), 132
Vicia villosa var. *villosa* (hairy vetch), 62, 88, 137
Viola adunca ssp. *adunca* (early blue violet), 33, 64, 113, 134, 138
Violaceae (violet family), 64, 113, 134
Viola glabella (stream violet), 64, 113, 134, 137
Viola howellii (Howell's violet), 134
Viola nuttallii var. *praemorsa* (upland yellow violet), 64, 113, 134
Viola praemorsa var. *praemorsa* (upland yellow violet), 64, 113, 134, 138
Violet Prairie, 6
volcanic ash layers, 8, 21
volunteers, 2, 26, 30, 36

W
wall bedstraw, 63, 104
Washington Department of Fish and Wildlife, viii, 6, 26, 29, 33, 35
Washington Department of Natural Resources, vii, 5, 14, 15, 20, 25, 27, 33
Washington Geological Survey, 51
Washington Territory, 43, 47-48
weed management, 38-39
western bittercress, 61, 78, 131, 143
western buttercup, 22, 33, 63, 98, 99, 133
western columbine, 56, 63, 97
western fescue, 135
western gray squirrel, 25, 67
western hemlock (western hemlock), 9-11
western pearly everlasting, 60, 71
western pearlwort, 132
western red cedar, 11
western white pine, 11
West Rocky Prairie, 5, 6, 36
wet prairies, 24, 26, 31
Whidbey Island, 47
white brodiaea, 64, 120

white fawn lily, 64, 118
white-flowered hawkweed, 130
white inside-out-flower, 131
White, Richard, 49, 50
white settlement, settlers, 12, 21, 25, 27, 43- 55, 57
white-stemmed lupine, 23, 62, 84
white-topped aster, *28*, 30-31, 35, 61, 75, 131, 136
wholeleaf saxifrage (swamp saxifrage), 63, 106, 134
wild carrot, 60, 68
wild cucumber, 61, 82, 132
wildlife sustained by prairies, 43. *See also* specific animals
wild strawberry, 32, 63, 101, 134
Wilkes, Charles, 45
Willamette Valley, 13, 24, 30, 32, 34, 46
willow family, 63, 105
witchgrass, 65, 127, 135
Wolf Haven, viii, 36
wood-sorrel family, 133
women, 17, 49
wood groundsel, 131
woods strawberry, 133
woolly sunflower (Oregon sunshine), 22-23, 29, 60, 72, 130, 136, 142
wormwood/sagebrush, 9, 10
Wyethia angustifolia (narrow-leaved mule's ears), 61, 76, 137

X
xeriphytic taxa, 10
Xerithermic, 10-11, 22

Y
yampah, 130
yarrow, 25, 33, 60, 70, 130
yellow and blue forget-me-not, 61, 77, 131
yellow glandweed, 64, 110, 134
yellow rocket, 131
yellow salsify, 61, 76, 131
yerba buena, 62, 91

Z
Zigadenus venenosus (meadow death-camas), 64, 120, 135, 137